Data and Knowledge in a Changing World

Springer

Berlin
Heidelberg
New York
Barcelona
Budapest
Hong Kong
London
Milan
Paris
Santa Clara
Singapore
Tokyo

J.-E. Dubois · N. Gershon (Eds.)

Industrial Information and Design Issues

With 102 Figures and 8 Tables

Springer

CODATA

CODATA Secretariat
Phyllis Glaeser, Exec. Director
51, Boulevard de Montmorency
F - 75016 Paris

Editors:

Professor Jacques-Emile Dubois
Université Denis Diderot Paris VII
Institut de Topologie et de
Dynamique des Systèmes
(ITODYS)
1 rue Guy de la Brosse
F - 75005 Paris

Dr. Nahum Gershon
The MITRE Corporation
7525 Colshire Drive
McLean, VA 22102
USA

The image on the front cover comes from an animation which shows worldwide Internet traffic. The color and height of the arcs between the countries encode the data-packet counts and destinations, while the "skyscraper" glyphs (or icons) encode total traffic volume at any site. This image was generated by Stephen G. Eick at the AT&T Bell Laboratories.

ISBN-13: 978-3-642-80288-1 e-ISBN-13: 978-3-642-80286-7
DOI: 10.1007/978-3-642-80286-7

Cataloging-in-publication Data applied for

Die Deutsche Bibiliothek – Cip-Einheitsaufnahme
Industrial information and design issues : with 8 tables / J.-E. Dubois ; N. Gershon (ed.). - Berlin ; Heidelberg ; New York ; Barcelona ; Budapest ; Hong Kong ; London ; Milan ; Paris ; Santa Clara ; Singapore ; Tokyo : Springer, 1996
(Data and knowledge in a changing world)

NE: Dubois, Jaques-Emile [Hrsg.]

Softcover reprint of the hardcover 1st edition 1996

SPIN:10542575 51/3020-5 4 3 2 1 0 - Printed on acid-free paper

INTRODUCTION TO THE SERIES

E. Fluck
Chairman of the Publication Advisory Board

CODATA's primary purpose in launching the Series "Data and Knowledge in a Changing World" is to collect from widely differing fields the wealth of information pertaining to the intelligent exploitation of data in the conduct of science and technology and to make that information available to a multidisciplinary community. This series, in support of that goal, will provide a forum for many types of submission, both theoretical and applied. Directories and Glossaries of compiled scientific and technical data or of Space and Time Observation data could be included in a separate **CODATA** Reference Data Series.

The present series will be wide in scope as it encompasses a broad range of data and knowledge contributions, including computer-related handling and visualization of data, to the major scientific, technical and medical fields.

To this end, the series on Data and Knowledge is open to contributions of various kinds, namely those:

- Fostering the improvement, not only of the quality and accessibility of quantitative and qualitative data, but of the classical and ground-breaking methods by which numeric and symbolic data are acquired, analyzed and managed;
- presenting new data and knowledge interfaces designed to optimize interoperability and thereby increase the potential of sharing data among databases and networks;
- intensifying international cooperation in communication and data sharing-this implies works dealing with standardization, data quality agreements and conceptual data descriptions (metadata, syntactic and semantic approaches)

along with papers dealing with the evolution of Internet-based facilities, other forms of worldwide communications and electronic publishing;

- providing new insights into, or interpretations of, processes leading to creative design in the field of concurrent and/or cooperative engineering, including cognitive aspects critical to data-based decision making.

In view of the emergence of the cyclic nature of information needs in the study of complex systems, **CODATA**'s scope now includes data activities and modeling of an interdisciplinary nature such as that encountered in far-reaching projects (e.g. Global Change, various Genome projects, environmental and biodiversity issues, etc.) or in certain medical information systems. Most of the above will therefore be highlighted in this Series.

Attempting, however, to predict the contents of such series is itself paradoxical, reflecting the particular challenge in the field of data and information management whose essence is to proceed at such pace as to outstrip the very information highways it is building.

In an evolving information world in which our traditional ways of transferring information as an essential resource are being altered, this Series should capture emerging and innovative concepts thereby leading ultimately to new information paradigms. Its ambition is to be concurrently a catalyst of change and a thought-provoking forum.

PREFACE

J.-E. DUBOIS and N. GERSHON

The first volume of this series, "**The Information Revolution: Impact on Science and Technology**", emphasized the importance of data sharing and fast communication and the advantages of current hypertext developments in creating new and flexible data access.

Volume II, "**Modeling Complex Data for Creating Information**", dealt, in particular, with the specific constraints of science and technology data including imprecision and uncertainty. It also provided representation and handling tools and object oriented programming technology for developing data systems.

The papers presented in this third volume are concerned with the very specific information problems of the technical and competitive industrial world. Here, production and selling rely on creative design, information processing, special up-to-date data search, knowledge comprehension and fast action, all essential for decision-making steps.

The following topics are discussed in this volume:

- **Cognition and Recognition in Design**
- **Knowledge Based Systems (KBS) Evaluation**
- **Modeling Tools for Knowledge Discovery**
- **Standards and CAD (Computer Aided Design) Aspects of Industrial Exchange and Specifications**
- **Information Seeking Strategies of Selective Access to Intelligent Information**
- **Special Information Resources: Complex Databases**

Most of these topics, inspired by the symposium on "**Communication and Computer Aided Systems**" held during the **14th International CODATA Conference**, deal with systemic components used by various up-to-date industries in development strategies. We see, for example, how STEP (Standard for the Exchange of Product Data) leads naturally to ISO (International Standard Organization) standards, allowing easier and more efficient data and information exchange. Several technical applications of STEP are presented.

The concept of life-cycle in industry is used here to introduce various important methodologies whose aim is to improve information sharing and transfer by standardized data supports. Faster intelligent documentation, implying quick information appraisal, coupled with massive on-line information servers, is a demanding goal that legitimates the presentation of several papers on textual analysis, graphics screening and visual data tools.

Information can be discovered and knowledge created from data sets by using modeling tools and artificial intelligence methods. One chapter is devoted to some important KBS modeling tools to manage numerical data and images in design processes.

Industrial developments often call upon expert systems or Data-Base Management Systems (DBMS) for production control, property evaluation or product design. In this book an entire chapter deals with the basic concepts of **CAD** and complex, often heuristic, **data-knowledge relations.**

In this third volume, the question of efficiency, implying rapid and easy information transfer, brings out the importance of the time element and of standards far more than in the other volumes. These factors are vital in the industrial world where one must ensure smooth flow and easy sharing of enhanced information.

These added conditions, felt less in academic life, justify the permanent quest for new tools to foster rapid communication between public and private information sources. This book provides exceptional insight into, and a global view of, the numerous facets of data building blocks and their role in industrial issue. It should prove a rich source of reference for engineers, computer experts and designers.

***Acknowledgments**: We wish to thank all the authors of this book for their contributions and hard work. We also wish to acknowledge the work and dedication of Bernice Dubois in helping to bring this book to the world and the tireless efforts and expertise of Belhadri Messabih in assembling the pieces from all over the world.*

CONTENTS

PART ONE. KNOWLEDGE: THE DRIVING FORCE IN CAD SYSTEMS

CHAPTER 1: CREATIVE DESIGN AND ITS COGNITIVE COMPONENTS INFORMATION

Chapter 2: Artificial Intelligence in Knowledge Based Systems (KBS)

Chapter 3: Modeling Tools for Knowledge Discovery

PART TWO. NEW DATA CHANNELS: INDUSTRIAL INFORMATION TOOLS

CHAPTER 4: INDUSTRIAL INFORMATION EXCHANGE: STEP, STANDARDS AND CAD DESIGN NEEDS

CHAPTER 5: INFORMATION SEEKING STRATEGIES: SELECTIVE ACCESS TO CURRENT INFORMATION ONLINE

Chapter 6: Information Resources : Intelligent Complex Databases

AUTHORS

Pascal ANDREI
UMLV/CESD, 2 rue Albert Einstein, 77420 Champs sur Marne, France

Alain BEZOS
Association GOSET, 107-111 Avenue Georges Clemenceau, 92000 Nanterre, France

Walter F. BOGAERTS
Katholieke Universiteit Leuven, Dept. of Materials Engineering, 3001 Leuven, Belgium

Hubert BRABANT
Laboratoire National d'Essais, 1 rue Gaston Boissier, 75015 Paris, France

Jean Pierre CALISTE
Laboratoire National d'Essais, 1 rue Gaston Boissier, 75015 Paris, France

Michel CARABEDIAN
Institut de Topologie et de Dynamique des Systèmes, associé au CNRS, Université PARIS VII,1 rue Guy de la Brosse, 75005 Paris, France

Marc CAVAZZA
C.E.S.D. - Thomson C.S.F., Laboratoire Central de Recherche, Domaine de Courbeville, 91404 Orsay Cedex,France

Aliette COSSE-BARBI
Institut de Topologie et de Dynamique des Systèmes, associé au CNRS, Université PARIS VII,1 rue Guy de la Brosse, 75005 Paris, France

Christine COUTRIS
Informatique CDC service RDT, 113 rue Jean-Marin Naudin, 92220 Bagneux, France

Anne-Françoise CUTTING-DECELLE
ESIGEC/LGCH, Campus Scientifique Savoie-Technolac, F73376 Le Bourget du Lac Cedex, France, E-mail: afcd@univ-savoie.fr

Dimitar N. DIMITROV
Bulgarian Academy of Sciences, 5000 Veliko Tarnovo, Bulgaria

Didier DONSEZ
Laboratoire MASI/UPMC, 4 Place Jussieu, 75252 Paris cedex 05, France
E-mail: donsez@masi.ibp.fr

Henri DOU
CEDOCAR, Faculté des Sciences de St Jérôme, Université d'Aix Marseille III, 13397 Marseille Cedex 13, France

Jean-Marie DOU
Bureau Japon, CEDOCAR, 26 Bd Victor, 00460 Armées, France

Marie-Françoise DOUTRE
Consultant expert, 101 rue Chardon Lagache, 75016 Paris, France

Vincent DOUZAL
Cemagref, BP 121, 92185 Antony Cedex, France

Anne-Marie DUBOIS
CSTB, BP 209, 06904 Sophia Antipolis Cedex, France
E-mail: anmarie@cstb.fr

Jacques-Emile DUBOIS
Institut de Topologie et de Dynamique des Systèmes de l'Université de Paris VII, associé au CNRS, 1 rue Guy de la Brosse, 75005 Paris, France

Jean FAGEBAUME
Laboratoire National d'Essais, 1 rue Gaston Boissier, 75015 Paris, France

Pascal FAUDEMAY
Laboratoire MASI/UPMC, 4 Place Jussieu, 75252 Paris cedex 05, France
E-mail: faudemay@masi.ibp.fr

Isabel GANCEDO
Ferroday Limited, 14 Mere Farm Road, Birkenhead L43 9TT, U.K.

Hélène HAON
Bureau Japon, CEDOCAR, 00460 Armées, France

Philippe HOMOND
Laboratoire MASI/UPMC, 4 Place Jussieu, 75252 Paris cedex 05, France
E-mail: homond@masi.ibp.fr

Leonid IANKO
International Atomic Energy Agency, Division of Nuclear Power, Vienna, Austria

Julian M. IVANOV
Higher Institute of Chemical Technology, 8010 Bourgas, Bulgaria

Shuichi IWATA
Research into Artifacts, Center for Engineering, The University of Tokyo, Komaba 4-6-1, Meguro-ku, Tokyo, 153, Japan
Faculty of Engineering, University of Tokyo, 7-3-1, Hongo, Bunkyo-ku, Tokyo, 113 Japan

Peter JOHNSON
Department of Computer Science, Queen Mary and Westfield College, University of London, U.K., E- mail: pete@dcs.qmw.ac.uk

Stoyan H. KARABUNARLIEV
Higher Institute of Chemical Technology, 8010 Bourgas, Bulgaria

Takashi KIRIYAMA
RACE, The University of Tokyo, Komaba 4-6-1, Meguro-ku, Tokyo 153, Japan
E-mail : kiriyama@race.u-tokyo.ac.jp

Christian KRUMEICH
C.E.S.D. - Madicia, Société Manos, 18, rue A. Einstein, 77420 Champs sur Marne, France

Uwe LANGBECKER
Technische Universität Berlin, Institut für Schiffs-und Meerestechnik, Salzufer 17-19, 10587 Berlin, Germany

Horris C. LEUNG
Warwick Manufacturing Group, Department of Engineering, University of Warwick, Coventry CV4 7AL, U.K.

André MARGUINAUD
Alcatel Espace, 5 rue Noël Pons, 92737 Nanterre Cedex, France

Ovanes G. MEKENYAN
Lake Superior Research Institute, University of Wisconsin, Superior, WI 54880, U.S.A

Belhadri MESSABIH
Institut de Topologie et de Dynamique des Systèmes de l'Université de Paris VII, associé au CNRS, 1, rue Guy de la Brosse, 75005 Paris, France

Naïma NAGUI-RAISS
PRISM Laboratory, Versailles University, 45, av. des Etats-Unis, 78000 Versailles, France, E-mail: naima@prism.uvsq.fr

Horst NOWACKI
Technische Universität Berlin, Institut für Schiffs-und Meerestechnik, Salzufer 17-19, 10587 Berlin, Germany

Gustaf ÖSTBERG
University of Lund, Box 118, S-221 00 LUND, Sweden

Clément PAOLI
Université de Marne-La-Vallée, Interface - 6, rue Alsace Lorraine, 76160 Darnétal, France

Ian R. PASHBY
Department of Engineering, University of Warwick, Coventry CV4 7AL, U.K.

Wenceslas PETIT
Université de Marne-La-Vallée, Interface - 6, rue Alsace Lorraine, 76160 Darnétal, France

René J. PISTENON
Institute for Research in Productics and Logistics, 24 Avenue Prieur de la Côte d'Or, 94114 Arcueil, France

Karin RABE
Department of Applied Physics, Yale University, New Haven, Connecticut 06520,U.S.A

Christo. REHLING
Technische Universität Berlin, Institut für Schiffs-und Meerestechnik, Salzufer 17-19, 10587 Berlin, Germany

John RODGERS
National Research Council of Canada, Ottawa KIA 0S2, Canada

Claude SENI
Atomic Energy of Canada, Ltd.,AECL,Mississauga,Ontario, Canada

Jean-Marie SÉZÉRAT
Informatique CDC service RDT, 113 rue Jean-Marin Naudin, 92220 Bagneux, France

F.J. SMITH
Artificial Intelligence and Applications Research Group Department of Computer Science, Queen's University, Belfast BT7 1NN, Northern Ireland, U.K.

Charles P. STURROCK
National Institute of Standards and Technology, Bldg. 223, Room B-254, Gaithersburg, Maryland 20899, U.S.A.

Norman SWINDELLS
Ferroday Ltd, 14 Mere Farm Road, Birkenhead L43 9TT U.K.

Gérard THIROT
Institut de Topologie et de Dynamique des Systèmes (ITODYS), Université Paris VII, 1 rue Guy de la Brosse, 75005 Paris, France

W. Y. TSUI
Department of Engineering, University of Warwick, Coventry CV4 7AL, U.K.

Edouard VALENSI
CEDOCAR, Centre de Documentation de l'Armement, 2 bis, rue Lucien Boussotrot, 00460 Paris Armées, France

Pierre VILLARS
Materials Phases Data System, Postal Box 1, CH-6354 Vitznau, Switzerland

Adam C WINSTANLEY
Artificial Intelligence and Applications Research Group Department of Computer Science, Queen's University, Belfast BT7 1NN, Northern Ireland, U.K.

Jianhua YAO
Laboratory of Computer Chemistry, Shanghai Institute of Organic Chemistry, Chinese Academy of Sciences, 345 Lingling Road, Shanghai 200032, China

Shengang YUAN
Laboratory of Computer Chemistry, Shanghai Institute of Organic Chemistry, Chinese Academy of Sciences, 345 Lingling Road, Shanghai 200032, China

Chongzhi ZHENG
Laboratory of Computer Chemistry, Shanghai Institute of Organic Chemistry, Chinese Academy of Sciences, 345 Lingling Road, Shanghai 200032, China

PART ONE

KNOWLEDGE: THE DRIVING FORCE IN CAD SYSTEMS

MODELS OF INFORMATION SYSTEMS: KNOWLEDGE, TASKS AND TOOLS

Peter JOHNSON

Department of Computer Science, Queen Mary and Westfield College, University of London, England, E- mail: pete@dcs.qmw.ac.uk

ABSTRACT

The thesis of this paper is that technology alone will not solve the problems of improving human abilities. The technology itself must be rooted in a sound understanding of the nature of human activity, to enable the technology to be developed to support and extend our knowledge rather than presenting hurdles and barriers to the development of human knowledge and ability. A tool that is not designed to complement and extend human knowledge and abilities is at best hard or impossible to use and at worst dangerous.

RESUME

Dans cet article, on avance l'idée que la technologie seule ne suffit pas pour améliorer les aptitudes humaines. La technologie elle-même doit être enracinée dans une saine compréhension de la nature des activités humaines pour permettre à la technologie le développement nécessaire pour supporter et étendre notre connaissance plutôt que de présenter des obstacles et des barrières à la progression de la connaissance et des possibilités humaines. Un outil qui n'est pas conçu pour compléter et étendre la connaissance et le savoir humains est au mieux difficile ou impossible à utiliser et au pire peut être dangereux.

1 INTRODUCTION

Most computer systems are intended to be tools for use by people in their work, leisure, and domestic activities. There are many different types of tools and computer technology provides a broad class of tools that aim to extend people's mental and in particular cognitive powers. Jerome Bruner postulated that the development of human knowledge evolves with the increased use of tools. The promise of computer technology is to provide powerful tools that can be effectively used in cognitive tasks. Through usage of these tools people should increase their knowledge and ability beyond their current limits. The reciprocal relationship between people's tasks, their knowledge and their use of tools is complex and crucial. The complexity lies in the nature of human knowledge, the tasks that people perform and the properties of computer tools. Moreover, the interactions between these three is also complex. How does human knowledge relate to tasks? What aspects of those tasks are tools required for? How do the tools relate to human knowledge? These questions have led to the development of an approach to the design of computer systems that places significant weight on modeling the knowledge and tasks of the people who will use the computer system as a tool in their activities. This approach is based on a theory of Task Knowledge [2] and subsequently an environment for developing computer systems. The thesis of this paper is that technology alone will not solve the problems of improving human abilities. The technology itself must be rooted in a sound understanding of the nature of human activity, to enable the technology to be developed to support and extend our knowledge rather than presenting hurdles and barriers to the development of human knowledge and ability. A tool that is not designed to complement and extend human knowledge and abilities is at best hard or impossible to use and at worst dangerous.

In the fictional country of Erehwon [1], Samuel Butler (1883) described a society in which people service the needs of machines rather than machines serving the needs of people and warned of the dangers of isolating considerations of users and the effects of technology on people's lives from the design and development of the technology. While in Erehwon the technology was that of steam engines and mechanical devices, the serious consequences of focusing design on the machine rather than on people and the device usage was clearly envisioned. In developing the technology of computer systems it is often necessary to focus upon properties of the technology. However, in developing systems that are intended to be used by people in the varied contexts of their work, private, social and leisure activities, the focus of design must be on the suitability of the designed artifact to support and complement human activity. Developing technology that serves the needs of people, rather than vice versa, requires a revised conception of computer system design that still allows systems to be well engineered so that they are easily maintained, reliable and efficient in their performance, but over and above this brings the needs of people and the context of usage sharply into focus in the design process.

An aim of the research we (and others) are engaged in is ultimately to develop methods and tools to bring about changes in the design process and, importantly, in the systems that are developed so that people and human activity will be better served by the technology of computer systems. The focus for our research has been to understand how knowledge of users and their tasks can be derived and incorporated into the design process. Our aim has been to develop an approach to design that places an understanding of people and tasks at the starting point of the design process, rather than solely as phenomena against which artifacts are evaluated.

In the ADEPT (Advanced Design Environment for Prototyping with Task models) project, we have developed an approach to design that includes both analysis and modeling of users and their tasks and the modeling, generation and refinement of interactive system designs from these user and task models. The ADEPT design environment provides a coherent approach to design which focuses the design around the tasks and activities of the users, rather than fitting the users and tasks to the design. This research differs from other recent work in developing advanced user interface design environments (for example see [4]) in that the ADEPT approach focuses on user tasks as the starting point for the design. Consequently, the designer is forced to recognize that design of an artifact entails the design of user tasks. In previous papers we have described the tools and models of ADEPT [3], [5] and [6]. The rest of this paper describes a generalized view of user-task centered design philosophy and then identifies where the models and tools of ADEPT fit within that view. This provides a useful way of identifying the strengths and weaknesses of ADEPT.

2 A User Task Design Philosophy

An approach to design which puts people and human activity at the fore of the development and deployment of technology should provide a framework for conceptualizing design and give rise to processes and tools that support that design approach. Figure 1 presents our conceptual framework for such an approach to design. In the first instance it relates people, tasks and organizations to artifacts, providing the basis for a design process that makes use of the relations. This generalized view leads to the development of techniques that allow information arising from users, tasks and organizations to be used in design.

People (as individuals and groups), tasks (as work, social, and leisure activities) and organizations provide the input to formulating the needs which should be satisfied by the design. This results in the development of what we have termed a designed task. The designed task is an expression of what an artifact will allow particular groups within known organizations to do and how they will be able to carry out those tasks, without prior commitment to the form of the technology or the detailed design of the artifact. This then leads to the design of an artifact (or collection of artifacts) that would enable the designed tasks to be performed. This Abstract Artifact Model is a description of the form of the artifact(s) without any commitment to concrete implementation detail. The Abstract Artifact Model is then refined into a Concrete Artifact Model which particularizes the design in an implementable form but still allows for alternative, platform specific versions of the artifact to be created that have the same physical and conceptual form. Throughout the development of the design of the artifact, the framework also provides for a re-expression of the task model that is expected to result from the further detailing of the artifact. Consequently, the design at all times is expressed in terms of the users, tasks, and organizations whose needs it is expected to serve.

To support this framework for design, our philosophy requires techniques that allow users, tasks, organizations and designers to participate jointly in the development of the artifact(s). This has caused us to make use of techniques from experimental psychology and anthropology that include role playing, observations, direct participation, interviews, card sorting, and protocol analysis. Similarly, the philosophy requires designers to ensure that people and organizations can directly understand how the artifacts resulting from design can be used. Consequently, facilitating communication between users and designers is a direct requirement on any method or tools that are developed to support this design philosophy.

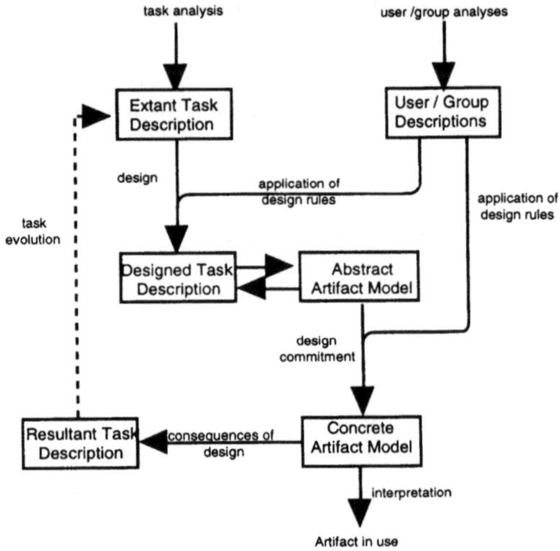

Figure 1. A framework for user task-based design.

A further concern is the evolutionary nature of design. There is an explicit recognition that while users, tasks and organizations must be strong determinants on the design of an artifact, the artifact will bring about changes to the needs of people and organizations and will give rise to new tasks. This is represented in the framework of Figure 1 by the connections between artifacts at the various stages of their design, and the resultant and extant expressions of the task consequences of the artifact. A final feature of the framework of Figure 1 required by the design philosophy is the need for the development of principles and other forms of design knowledge that support the transitions between the needs and characteristics of users, tasks and organizations and the properties and functions of artifacts.

In summary, the design philosophy assumes that:

- Good user interface design comes about through having increased involvement, communication and participation in design by designers, users and organizations;
- Increased involvement and communication can come about through greater understanding of users, their tasks and organizations and the consequences of usage of the artifacts;

- Design must include a focus on the tasks and activities of the users and organizations;
- It should be possible to ignore and abstract away from the detail of implementation issues, i.e. not force over-commitment in design;
- The design process requires iteration around the development of all models including those of the users, organizations, tasks and the artifacts;
- A final assumption is that design requires the development of principled design knowledge to support the transition between user, organization and task characteristics on the one hand, and artifact properties and functions on the other.

In the subsequent sections of this paper the ADEPT design environment is described relative to this design philosophy and the framework. A fuller description of ADEPT has been given in [5] and [6].

The models and tools of ADEPT have been developed in line with the above design philosophy. However, ADEPT is not necessarily the idealized or optimum environment for such a design philosophy. It is our first significant attempt at developing tools to support such a view of design. Furthermore ADEPT can be used as part of many different design methodologies that take seriously the importance of user and task modeling as an input to the design of a computer system and its user interface. The ADEPT tools assume that a task analysis and a user group characterization has been performed. A comprehensive method of task analysis that produces a rich input to the form of TKS models of ADEPT is described in [2]. The user characterization can be carried out at the same time as the task analysis using questionnaire, interviewing or self reporting techniques. The aim of the ADEPT models and tools are to make the designer think explicitly about the user group and the tasks or purposes to which any design should be put. The designer is required to explicitly model the nature of the work of the user organization in terms of the roles and tasks that form the substance of that work. The designer is further required to explicitly model how the design will be used as part of that work or how it will be used as part of some envisaged changed work situation. Through using this approach the designer is able to obtain an understanding of the users and their work and the ways that any proposed design options might fit within or change that work context. As part of the process of design it is important that the designer and users are able to communicate about the design and its context of use. To facilitate this communication it is important that the users have a format in which they can describe their tasks and which can be understood by the designers. This communication results in the task models of the ADEPT environment being constructed. Similarly it is equally important that the designer has a format through which he can relate his design ideas to the users thus enabling the users to see how the design relates to their work tasks. Through the use of the ADEPT models the users and designers are better able to communicate with each other about their relative areas of expertise - namely work and technology.

The models and tools of ADEPT have been developed from the above design assumptions of a task based approach to design that allows user and designer participation in the design activity. It provides a focus on design issues that affect user tasks and user activity while allowing design considerations resulting from the details of the implementation to be also considered without deflecting the designer away from the important and often overlooked aspects of the context of usage of the design.

3 THE ADEPT DESIGN METHOD

The first point to be made about the ADEPT method is that it is not a proceduralized method of design that dictates the steps to be gone through in producing a design. Design is not like that: it cannot be proceduralized. Design involves problem solving and the models and tools of ADEPT facilitate identification, definition and refinement of the problem space and development and evaluation of the design solutions. The ADEPT method can be closely associated with rapid prototyping approaches to design and scenario and participatory approaches to design.

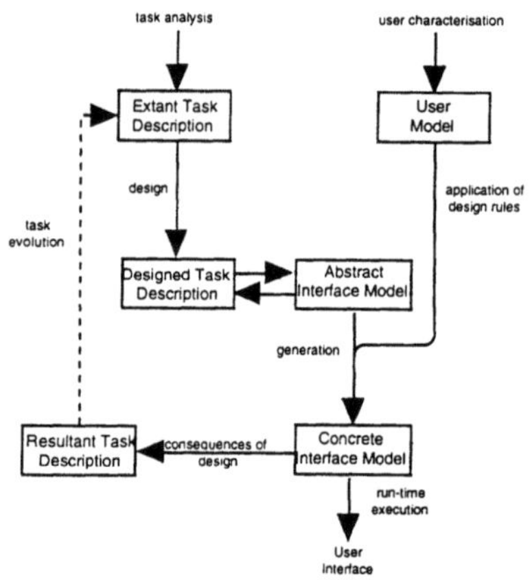

Figure 2. The ADEPT design process.

ADEPT user and task descriptions. The task and user descriptions of ADEPT form an important input to the definition of the design problem space. The task model is a detailed description of the roles and tasks that are currently performed as part of some work activity. The task model details the goal structure and the procedures by which these many and varied goals may be carried out. This model constitutes a description of the existing work context from the perspective of the users' goals and tasks, according to the various roles that are taken on. The user model is a description of the characteristics of the individual and/or groups of people who commonly perform these defined roles and their associated tasks. As such together they can be thought of as components in the definition of the requirements for the design. More correctly, they constitute a definition of the context into which a design would have to be introduced. The form of the description of those requirements is not constrained in any way by ADEPT other than in terms of their content. The content must make explicit the details of the tasks and users' characteristics as understood by the designers.

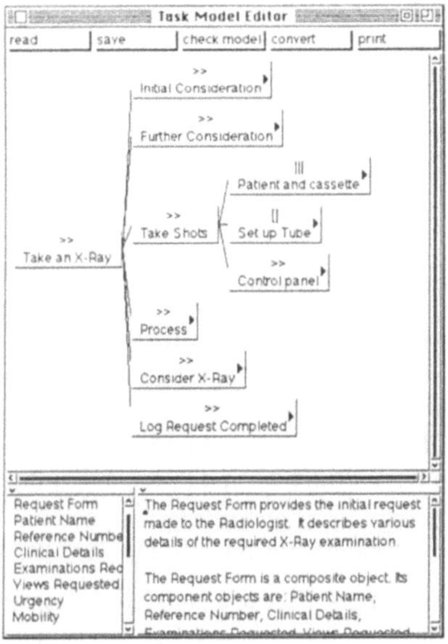

Figure 3. ADEPT Task Model Editor.

The techniques used to develop the task and user models include methods from experimental and social psychology and anthropology. The models are developed first and foremost by the users themselves with the help of the designer/analyst in providing summative descriptions of the information gathered which the users can check and change. The techniques include participation in the user tasks by the designer/analyst and role playing by the users and designer/analyst. Indirect, non-obtrusive observations are carried out as well as collecting users' own reports and descriptions of their tasks. For a fuller description of the techniques used see [2].

The nature of the task and user models in ADEPT are such that they have different forms of presentation. The task model is presented as a Goal and Procedural structure diagram and a Task Domain Object list with textual descriptions linking the objects to the actions of the tasks and to other objects. The User Model is presented as a list of rules. The descriptive task and user models represent the working material for the progression of the design from which a more formalized (i.e. standardized) textual requirements definition may be produced to summarize this information. Where necessary, the relevant extracts from the task and user models can be inserted in the requirement documents to provide direct links between the models and the resultant requirements.

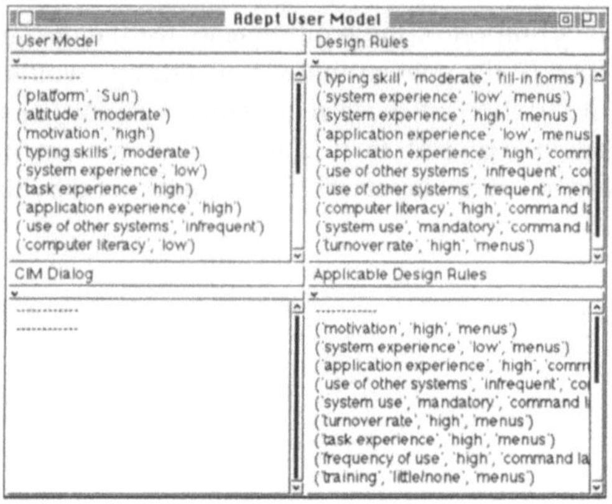

Figure 4. ADEPT User Model.

Standard approaches to software design such as SSADM (Structured Systems Analysis and Design Method) and JSD (Jackson System Development) assume that requirements will be written in a natural language and will largely be of a functional nature. The focus of the ADEPT approach is to add to these requirements detailed descriptions of the user and task context into which the design must fit.

Decompositional Design - design task model, AIM and CIM

ADEPT does not follow a standard waterfall model for design which assumes that all requirements can be elicited before the start of design and that high-level design can be completed before any low level design. However, ADEPT does force the designer to start from users and tasks and to progress to both high and low level design issues. It is difficult to envisage a situation in which the design of a system should progress from an artifact to a search for its possible uses. (Note, technology may progress in this fashion on some occasions but the development of technology is not the development of artifacts. Ideally, artifact design progresses using available technology or creates a need for the development of new technology.) The ADEPT tools support the designer in constructing a task model of the design. This is a model of the tasks that the designer envisages the user carrying out with the designed artifact. This task model may be a subset of the existing task model (Fig. 5B). Alternatively, the design task model may be an overlapping extension of the existing task model (Fig. 5A) allowing the user group to carry out tasks and attain goals which were not previously part of their responsibility (either because they could not be carried out or because they were carried out by different people or agents). A further option for the design task model is for it to be simply a revision of the existing work by changing the way that the existing tasks are performed (Fig. 5C).

It is clearly impossible for the designer or the users to conceive of all possible tasks in all possible contexts to which the design might be applied. Instead, the design task model is a description of those tasks which the users and designers agree that the system must be able to support. The design task model is then the high level specification of what the design will allow the user to do. Thus it is the first attempt at meeting the design requirements by saying what the computer system will enable the users to do and how it will affect their work. The designer can/should evaluate this stage of the design by comparing it with the existing task model for its scope and its consequences on the work and tasks of the users.

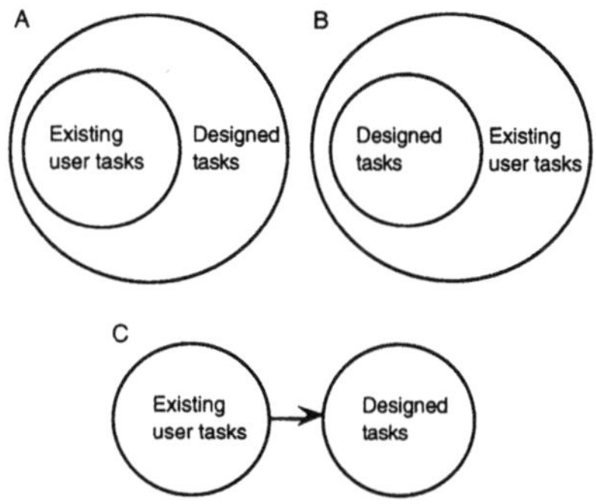

Figure 5. Possible relations between existing tasks and design: A shows design extending the range of tasks users can perform, B shows design supporting a subset of existing tasks and C shows design replicating current tasks with a change of technology.

From the design task model the designer can produce the Abstract Interface Model (AIM). The AIM is a model of the interaction aspects of the proposed design. This model details the dialogue that the users will engage in to carry out any of the proposed tasks of the design task model. In addition, the AIM details the designed objects that the user will interact with in the course of these dialogues. The relationship between a task, an interaction dialogue and the interaction objects is the focus of this model. The dialogue model is defined in terms of the designed tasks. The completeness of the dialogue model is an issue of concern for the designer. The dialogue model is as complete as the design task model in that it must describe how each task of the design task model would be carried out by the user(s). It must contain sufficient detail that the designer can be sure that each step in an interaction is modeled. This represents the idealized dialogue for using the design to carry out each of the defined tasks. In addition to this, the dialogue model of the AIM contains a state-event representation of what states can be reached from other states so as to allow the designer to identify any deadlocks, etc. in the design. The designer and the users are able to "walk through" the dialogues as if they were interaction scenarios for particular tasks. In this way the designer can involve the user in the evaluation of the design at this early stage.

The object component of the AIM provides a complete description of all the objects necessary to support the designed dialogues and to allow the designed tasks to be carried out. This is a textual description of the objects, their properties and the links to other objects including dialogue links. The presentation format of the AIM is similar to that of the Design Task Model and the Extant Task Model. This is to enable both the designer and the user to readily relate each of the models to each other and thereby ensure that the design evolves with its envisioned usage most prominent.

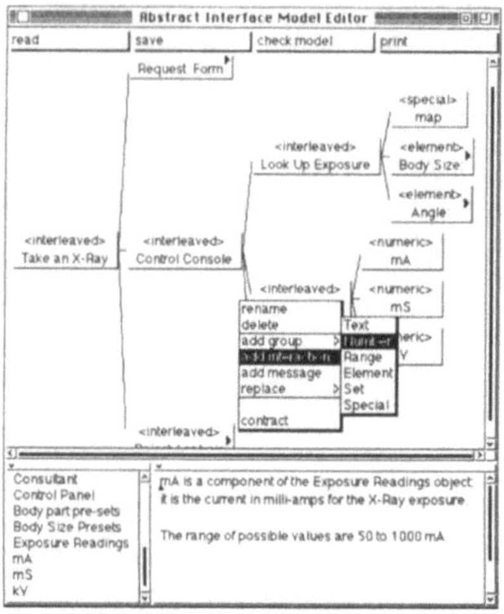

Figure 6. ADEPT Abstract Interface Model.

The AIM can be developed further to produce a Concrete Interface Model (CIM). The method by which the CIM is developed is through refining the design to detail how the interaction described in the AIM is to be realized through an actual computer interface. This includes details of the interaction style to be used and the form of the interaction behaviors of the user and the interface. In this model the designer must define when the user will provide an input or receive an output from the computer in terms of the medium and modality through which that input and output will occur.

The CIM must determine for example, when keyboard, mouse, pointing, voice or other form of input will occur, and when graphical, textual, sound or other forms of output will occur. The ADEPT tools present the results of these design decisions in the form of a runnable, skeletal model of the interface. As with all the other models of ADEPT, the designer and the user can understand the consequences of the design for the users' tasks. Through inspecting and running the CIM the user and designer can begin to assess what it will be like to interact with the computer to carry out the designed tasks, or other tasks that were not included as part of the original designed tasks. From the CIM the designer can generate a runtime interface on a particular implementation platform. The result is a complete system (in the sense that it has all the physical and behavioral forms that have been so far designed).

Throughout the design life-cycle the ADEPT models and tools focus on users and tasks and stress the need for user and designer involvement in a participatory design process. The need for iteration and participation of users and designers is throughout. The task and user analyses must be iterative and participative as is the development of the task model, user model, design task model, abstract interaction model, and concrete interaction model. The designer and user can view the design in relation to its usage in task contexts. At any stage in the design the user or designer can reflect on the adequacy of the design, in terms of its scope, dialogue, objects, or the style and form of interaction. For example, it is expected that as designers and users advance the design, users will identify new tasks or ways of performing extant tasks that were not included in either the extant task model or the design task model. ADEPT will allow the designer to then go back and redevelop the relevant model. This is not an inadequacy of any model or analysis technique; it is a realization that the design environment must allow for iteration at all stages of design and to allow for existing and future task contexts to be modeled as part of the design process.

With such an approach to design, evaluation is no longer a separate activity that is carried out at discrete points in design. Instead it is a central feature of the iterative process that occurs every time a user or a designer reflects upon a given model or set of models. ADEPT supports both formative and summative evaluation. It places great emphasis on evaluating the models directly with the users, and also upon the evaluation of all design decisions in terms of their effects on the usage of the system in the context of work tasks.

CONCLUSION

In the framework for user task-based design, we identified task, user and organizational descriptions as constituting important inputs to the design process. The ADEPT tools support particular forms of task and user descriptions, but lack any organizational description. The particular forms of task and user description are highly manipulable but are relatively formal in their appearance. The task model is a combination of a graph diagram of goals and procedures and a textual description of objects, while the user model is a purely rule-based description. The form of the user model leaves much to be desired both in terms of the content of the rules and in its ability to support user participation in design. The designed task description is an important feature of the design framework and is present in ADEPT but is not well supported in providing the designer or the user with any form of rationalization or argumentation structure that might be used to support particular reasons for changing or preserving the form of the tasks. In addition, there is little in the way of design principles or design knowledge that either the user or the designer can call upon to support their decision making. The Abstract Artifact Model of the framework is exemplified by the Abstract Interface Model in ADEPT. The AIM is a clear description of the intended design without forcing the designer to commit to particular design detail over the implementation. However, it is not directly executable or interactive other than in its construction, consequently neither user nor designer can begin to use the design at this stage. The Concrete Interface Model is a fairly accurate realization of what a Concrete Artifact Model should include. It is interactive and executable. However, it is not easily modeled by the user so the designer is likely to have control over changing the CIM in ADEPT. Our aim will be to make this an operation which the user can also perform so that both designer and users can contribute directly to the development of the detail as well as the abstract form of the design.

It will be clear from the preceding discussion that the ADEPT design tools have evolved from our philosophy yet lag behind our understanding of the problems in carrying out task-based design. We are concerned that there is a lack of design principles which can be used to guide the designer in developing applications from the perspective of users and their tasks.

ACKNOWLEDGMENTS

The Adept Project (Advanced Design Environments for Prototyping with Task Models) was a collaboration between Queen Mary and Westfield College (University of London), British Aerospace Plc, British Maritime Technology Ltd and MJC2. It was led by Queen Mary and Westfield College and funded by the DTI and SERC (Grant No. IED 4/1/1573). Current work is proceeding under direct contract funding to QMW from British Aerospace Plc.

KEYWORDS

Abstract artifact model, ADEPT design, interface philosophy, human knowledge

REFERENCES

[1] Butler S., A. Erehwon, Routledge, London (1883)

[2] Johnson P., Human-Computer Interaction; Psychology, Task analysis and Software Engineering. McGraw Hill, Maidenhead (1993)

[3] Johnson P., Wilson S., Markopoulos P., Pycock J., ADEPT - Advanced Design Environment for Prototyping with Task Models, In INTERCHI'93 Proceedings of Human Factors in Computing, Amsterdam, Addison Wesley, ACM press, p.56 (1993)

[4] Neches R., Foley J., Szekely P., Sukaviriya P., Luo P., Kovacevi S., Hudson S., Knowledgeable Development Environments using Shared Design Models, Proceedings of 1993 International Workshop on Intelligent User Interface, ACM press, pp. 63-70 (1993)

[5] Wilson S., Johnson P., Colgan L., Kelly C., Cunningham J., Beyond hacking: a model based approach to user interface design, In Proceedings of BCS HCI93, Loughborough, Cambridge University Press (1993)

[6] Wilson S., Markopoulos P., Pycock J., Johnson P., Modeling Perspectives in User Interface Design, Proceedings of East-West Conference on Human-Computer Interaction, pp. 210-216 (1992)

On The Management of Incompatibilities in Knowledge Systems

Gustaf ÖSTBERG

University of Lund, Box 118, S-221 00 LUND, Sweden

Abstract

In our traditional school education we are induced to think that we can apply systems concepts when dealing with problems not only in mathematics, physics and chemistry but in other fields also. The success of the systems approach in these sciences makes us believe that the same methods can also be satisfactorily applied to live organisms. The failure of systems analysis to account for the behavior of such organisms is often blamed on deficiencies in the logic of the systems. It appears, however, that the main reason for the lack of success of attempts to develop, for instance, computerized data bases for materials to be used by designers is that they have neglected other factors than the hard, quantifiable ones. This is particularly true for tacit or personal knowledge and unknowable factors. In order to cope with these deficiencies and the incoherence in so-called natural systems one has to consider them in their own context. Even if most incompatibilities can then be overcome by means of management, there may remain value conflicts that cannot be settled except by the use of power.

Resume

Dans notre système d'éducation traditionnelle on est enclin à penser qu'il est possible d'utiliser les concepts des systèmes dans tous les domaines et pas seulement en physique, chimie et mathématiques. Le succès de l'approche systémique dans ces sciences conduit à penser qu'elle peut être appliquée de façon satisfaisante aux organismes vivants. L'échec de ces analyses pour représenter le comportement de tels organismes est souvent imputé à des défauts de logique de ces systèmes. En fait, la raison principale de nos insuccès à développer, par exemple, des bases de données informatisées pour les matériaux à utiliser par les "designers" (concepteurs), est liée à l'oubli de certains facteurs autre que les facteurs essentiels bien quantifiables. Pour faire face à ces défauts et à l'incohérence des systèmes dits "naturels", il est nécessaire de les aborder dans leur propre contexte. Même si la plupart des incompatibilités peuvent être surmontées par le biais d'une certaine organisation, certains conflits de valeur ne peuvent être réglés autrement que par une décision autoritaire.

1 Introduction

Schematically the nature of this paper is descriptive and explorative rather than normative. It deals with experiences of some systems which are usually classified as technical, namely safety systems for nuclear power and data systems for materials. The point to be made, however, is that such systems are often dealt with in ways similar to those employed for mechanical or hydraulic machines. Differences in character between different components are neglected and less visible; non-technical factors are ignored.

Since these deficiencies in the account of such systems may be critical for the understanding of them, an interpretation of the incompatibilities in question will be given with a view of providing means for handling these systems. Admittedly, the description and explanation of the problem addressed are made by a novice in systems theory. It is the stimulation given by colleagues specializing in more rigorous theoretical treatment of systems that is the author's excuse for presenting the following views to an international audience and inviting its criticism.

2 Incompatibilities in Physics

The meaning of the word incompatibility in this context is probably best illustrated by examples from basic mathematics and physics. At school, when we learn to calculate the circumference of a circle, we are told to multiply the length of the diameter by π . At that stage most teachers, if not all, hesitate to mention the fact that the result of this calculation is an irrational or transcendental number. This is, of course, contrary to what people consider the physical reality, i.e. that the circumference is an exact measure.

A similar example is the calculation of the hypotenuse of a right angled triangle according to Pythagoras. Assuming the length of the smaller sides to be 1, the result is again an irrational number. To take yet another case, in the physics of crystals, distances between atoms occupying fixed positions in a space lattice are in principle assumed by most physicists to be exact numbers, although by calculation they arrive at irrational, indefinite and infinite numbers.

For the purpose of this paper the lesson of these considerations is psychological. We become used from childhood to accepting, without very much reflection, fundamental discrepancies between our physical concepts on the one hand and our mathematical models on the other hand. Constants like π are employed to make it possible to equate phenomena or factors, in order to make it possible to describe and handle relationships in nature.

In our everyday life cases like those just mentioned are of little practical importance. Modeling without proper consideration of possible gaps in our understanding of the differences in the character of phenomena can become serious the more unknowable factors our systems contain. This seems to be true for materials in particular.

Materials are made up of atoms which are mostly arranged in symmetric patterns, crystals and molecules. The latter are joined in structures of increasing order until we reach the scale we can observe with the naked eye. For each level - atoms, crystals or molecules, etc. models and equations have been developed to describe and explain the properties of these respective aggregates. What is important to note is the fact that these models for each level are not mathematically compatible to the extent that would enable us to calculate the performance of a certain material in, for instance, an engine, solely on the basis of the properties of atoms, crystals or molecules.

Only in such mechanical or electrical devices where the material operates within a limited range with respect to structural complexity can its performance be predicted from structure and chemical composition. There is one classical example of this, the transistor. For components or products with a higher degree of structural complexity a designer has to rely to a large extent on experience in order to overcome the gaps in the total mathematical model of his materials.

This reliance on experience may be possible as long as the systems are transparent, i.e. one can recognize the functioning of and relationships between the different parts of the system. With increasing abstractness of industrial and societal systems this recognition becomes more and more difficult. As a consequence we have to consider the need for limiting the complexity of our systems before we reach the stage where we no longer know whether we can control them to the desired degree.

3 Design Mechanisms

The physical aspects of materials are only one set of coordinates where incompatibilities may occur. A designer has to consider materials from several different points of view apart from basic properties: availability, aesthetics, recoverability to mention just a few. Obviously, he has to deal with a conglomerate of sub-systems between which the links are only weak, if indeed they can ever be identified.

At this stage it should be quite natural to take a step backwards in order to find out how designers manage to do their work in practice. Apparently they mostly seem to be able to handle the complexities and incompatibilities involved in quite a satisfactory way. Furthermore this cannot always throughout be a trial-and-error process.

In case design really means solving problems of incompatibility in complex systems the designer's trick is probably to treat these problems on other levels than those where the incompatible phenomena occur and meet. This was what Pythagoras did when he solved the problem of irrational numbers by moving from the linear dimension of the sides of the triangle to their quadrates. In industry this is what the general manager does when he is faced with incompatible and often contradictory opinions from the different parties and departments of his company. In this case regarding the coordination problem from a point outside is then formally defined as applying the business idea of the enterprise.

Even then, by reference to a certain business idea, the manager or the designer may be left with disagreements which cannot be solved, at least not at a particular moment in time. The process of management or design is a dynamic one which means that any decision has to be based on a judgment of the likely development of the balance between different interests and parties. Ultimately this is a matter of power.

The judgment of a manager or a designer must then be based on an estimate of the interests of those at whom the activities are aimed. In the case of the safety of nuclear power, which is one problem area considered by the author, the interest of the end-user is not simply cheap and reliable electricity. If the generation of electricity by means of nuclear power does not meet the public's standards of safety the normal technical performance of the nuclear reactors is of less interest.

The same is true for data systems for materials. However comprehensive the system from the computer and materials science point of view, it is useless and worthless if it does not satisfy the designers' needs for knowledge of the technical requirements they must consider.

4 A Question of Interest

When the issue is stated as simply as that, most systems designers find it easy to agree about the need for appreciation of non-technical factors.

Experience says that in practice very few technologists pay more than lip service to designing systems according to the paradigms of their own trade. There is thus in the end an overwhelming problem to find a way to make hard systems designers also respect the soft aspects in their own right.

There is nowadays a growing insight among psychologists that our perception of a technological system is related to some mental model of how the system should naturally be constructed. The mental model may differ from the artificial one in order to make the maximum possible use of the inherent capability of systems design. It is the discrepancies between these models that are the cause of the malfunctioning, or the nonfunctioning, of some systems for nuclear safety and data systems for materials respectively.

Formulated in these terms the lack of respect for soft factors can be traced back to mentalities and attitudes. It is the author's opinion that not much can be achieved towards making complex systems efficient until a dialogue can be established between systems designers and users on an equal footing. This finally boils down to an exchange of views on the goals of systems design, taking into account both parties' world picture.

5 CONTEXTUAL COMMUNICATION

Coordination of seemingly incompatible activities in an organization by the application of common ideas may be called management by objectives. It should be noticed, however, that simply stating objectives or goals from a manager's point of view does not necessarily solve the problem of matching differing factors in the system. Experience says that a message about objectives articulated by management is often not perceived at all by the audience. It escapes those for whom it is intended because it is constructed on the basis of concepts that seem irrelevant to people outside the management's sphere.

This dilemma of communication is essentially not a question of language, expressions or wording. Neither can it be dismissed as a trivial symptom of bureaucratization which could be remedied by using organizational methods. Its origin can be understood only by consideration of ends versus means. While it is true that ends and means are not necessarily independent of each other these two concepts might be identified as phenomena of different natures for the purpose of the present discussion.

The difference between ends and means from the management point of view can be characterized as a vertical incompatibility as compared to the horizontal incompatibilities dealt with above. In this case the management itself is one part of the incompatibility. Hence it cannot resolve the problem by adopting a view from outside, while at the same time remaining inside.

Since the crucial point of the dilemma in question seems to be an inability on the part of the organization to perceive and comprehend other messages than those referring to the means of their activities, the communication has to be based on references to these means. In order to be perceived, ends have to become subordinated to means in appeals from management. Ends have to be dealt with from the context of means.

At first glance this appears to be nothing else than the well-known rule that in order to captivate an audience a speaker should make reference to the personal interests of the listeners. There is an analogy, however, which indicates that the process of perceiving an issue on a higher level of abstraction requires a mutual understanding on the parts of both parties involved. In the science of literature a certain process called the hermeneutic circle is recognized. In order to understand the full meaning of a piece of literature one has to see not only its broad outlines. It is necessary also to examine the details of the wording. Conversely, the latter cannot be very well understood without the background of an interpretation of the overall meaning. Hence the process of understanding literature is a circular or spiral movement.

It appears as if ignorance of the hermeneutic circle could explain the lack of success in communicating messages about ends to people who are assigned to confine themselves to means within an organization. On the other hand, successful recognition of the interdependence of ends and means might eventually lead to a holographic effect. By this is meant that the whole can be perceived from observation of only a fragment according to the principles of optical holograms. Admittedly, however, this would require that both ends and means have something in common, which may not necessarily always be true or desirable.

Obviously, basing an implementation of ends on the grounds of means involves risk of trivialization of the ends. In this process a manager will easily find himself in a neurotic state. This is no indication of mismanagement. It only shows that management has to do with man-made systems.

Conclusion

As Kierkegaard says: "A logical system is possible; an existential system is not possible".

Keywords

Contextual communication, design, ends and means, hermeneutic circle, holographic, incompatibilities, interest, irrational numbers, knowledge systems, unknowable

Reference

Kierkegaard S., Unscientific Postscript to the Philosophical Fragments, A Mimic-Pathetic-Dialectic Composition, an Existential Contribution By Johannes Climacus (1846)

KNOWLEDGE BASES FOR CREATIVE DESIGN OF MICROMECHANISMS

Takashi KIRIYAMA

RACE, The University of Tokyo, Komaba 4-6-1, Meguro-ku, Tokyo 153, Japan
E-mail : kiriyama@race.u-tokyo.ac.jp

ABSTRACT

Micromechanism design is an interesting and challenging field for design engineering. Its interdisciplinary and innovative characteristics are used to investigate the possibilities of supporting the design process with knowledge based systems. This paper presents a design environment for the conceptual design of micromechanisms. The environment has been implemented on a knowledge based system SYSFUND. Knowledge bases about functional primitives, physical features, and physical laws have been built in the design environment. The use of these knowledge bases is illustrated by an example of micro accelerometer design.

RESUME

La conception de micromécanisme est un champ d'activité intéressant et un défi pour l'ingénierie de conception. Ses caractéristiques interdisciplinaires et innovatrices sont abordées en vue d'étudier les possibilités de soutenir le processus conceptuel avec des systèmes à base de connaissances (KBS). Cet article présente un environnement de design pour la conception de micromécanismes. L'environnement est implémenté sur un KBS nommée SYSFUND. Les bases de connaissances (KB) relatives aux primitives fonctionnelles, aux faits physiques et aux lois physiques sont construites dans cet environnement. L'utilisation des KB est illustrée par un exemple visant à définir un microaccéléromètre.

1 INTRODUCTION

Micromachine technology has developed significantly in recent years and become an outstanding research field with remarkable results. In the past several new products resulted from this field, mainly related to sensors, actuators, and medical apparatus. The characteristics of micromachines are given by the interaction of miniaturized mechanical and electronic components, which are manufactured by methods originally developed for semiconductor production. These characteristics of micromachines are used to investigate the possibilities of supporting the micromachine design process with knowledge based and design methodological techniques.

The requirement for the support of this design process by adequate design systems can be derived from the importance of micromachines for the development of future products. Today's design systems, as well commercial systems as scientific prototype systems, show more or less increasing capabilities of supporting the conceptual design stages, which are important for the micromachine design process [1-3].

However, many questions are still open and there is but little experience available as to how today's concepts for design systems can support micromachine design. The main intention of this work is to investigate the possibilities for supporting the micromachine design process by so-called knowledge based design systems.

This paper describes the development of a design environment for micromechanisms. The role of the design environment is to provide the designer with useful information to explore creative design concepts.

Several kinds of knowledge bases are built in the design environment. The knowledge base about functional primitives is used to search for possible solution patterns in the conceptual design stage. The knowledge base about physical phenomena helps the designer to embody the functional hierarchy as the physical structure. The knowledge base about physical laws helps rational thinking about scaling effects. A knowledge base about fabrication processes helps a group of designers to share the knowledge about available manufacturing technologies over the network. These knowledge bases are described in the rest of this paper.

2 KNOWLEDGE BASES FOR MICROMECHANISM DESIGN

In building the design environment for micromechanisms, a knowledge based design system called SYSFUND has been used. SYSFUND (Systematization tool for functional design and synthesis) [4] is a software for building functional and physical knowledge bases for the use of modeling and reasoning about the design object. SYSFUND uses a modeling framework called the FBS (Function, Behavior and Structure) modeling. In the user interface of SYSFUND, an FBS model is presented as a graph of components of the function, behavior and structure. The user can build an FBS model by combining these components which are stored in the system databases.

The functional part of the FBS model represents how an abstract function is hierarchically decomposed into more concrete sub-functions. A functional component consists of a verb and an object which receives the effect of the function (i.e. to do something to an object). It may take another object by means of which the function is performed (i.e. to do something to an object by means of another object). In building a functional hierarchy, the user may ask SYSFUND to suggest possible decomposition of a functional component into sub-functions. If there is some known functional decomposition stored in the database, the user can select one.

The behavioral and structural part of the FBS model represents how the functional structure is physically embodied. This part is represented as a causal network of physical phenomena and structural elements, in which a physical phenomenon is connected with prerequisite physical phenomena by causal links. The user constructs the behavioral and structural part of an FBS model by retrieving so-called physical features associated with functional components. A physical feature is a chunk of physical phenomena and related structural elements.

3 ENHANCEMENT TO SYSFUND

Although SYSFUND provides a basic framework of modeling mechanical systems, for a specific class of design problem its knowledge bases and tools must be augmented. In order to adapt SYSFUND to the design of micro-accelerometers, the features of SYSFUND were enhanced as described below.

Representation of functional components:

It was decided that functional components would be represented as input to output transformations. This means that all functional components are defined by a pair of input and output quantities, i.e. to transform the input to the output.

Although this is restrictive compared to a more general definition to do something to an object, originally employed by SYSFUND, for the design of micro-sensors and actuators, it was assumed that this representation covers a wide range of known functional components for the design of micromachines. In addition, this change made available the next functional reasoning capability.

Functional reasoning:

SYSFUND does not automatically search for sub-functions that have the same effect as the overall function. But since functional components are represented as input to output transformations, the equivalence between a functional component and a sequence of functional components was defined simply as the equivalence between their inputs and outputs.

Based on this definition, a search algorithm was implemented to detect all possible sequences of functional components that have the same input and output quantities as specified functional components. For an overall function if some sequences of components of sub-functions are found to be equivalent, they are used as candidates for functional decomposition.

Scale analysis:

An external module of scale analysis was added to SYSFUND. This module is intended to help the designer to identify physical phenomena effective for the design of micromachines. Equations for the scale analysis are stored in the physical feature database associated with each physical feature. For instance, with a physical feature "bending of a beam" the following equation is associated.

$$v = \frac{3pl^4}{2Eh^2} a$$

In the equation, x is the displacement at the end of the beam of density r, length l, Young's module E, and thickness h under a perpendicular acceleration a. The scale analysis module allows the user to visually examine the scaling-down effect on physical quantities. It displays the magnitude of each parameter graphically. For instance, according to the equation, the displacement at the end of the beam x decreases to the fourth power the length of the beam l. If the user selects x as the target parameter and reduces l on the display, x decreases four times faster than l. If thickness h is reduced in proportion to the length l, the displacement x decreases to the second power. This analysis shows that the relative distortion of a micro scale structure is much smaller than that of an ordinal scale structure, and therefore a cantilever for detecting acceleration must be made very thin to be sensitive enough.

4 MODELING THE TUNNELING CURRENT ACCELEROMETER

In the enhanced SYSFUND, necessary knowledge for the design of accelerometers was added, including functional components, physical features, their correspondences and physical laws. By using these knowledge bases, a design process of a micro accelerometer was performed. This device, measuring less than a square millimeter, has a cantilever which deflects when the device is accelerated. The deflection causes a change in the gap between the end of the cantilever and a fixed electrode. A tunneling current across the gap varies depending on the gap, and thus the acceleration of the device is measured.

Figure 1 shows a screen copy of the FBS model of the tunneling current accelerometer. In the white upper window the functional hierarchy is presented, and in the gray lower window physical features are arranged. A list on the right shows available functional components. Vertical links across the windows connect corresponding functional primitives and physical features.

For the main part, the device's function: to detect acceleration, functional decomposition was performed. In the functional hierarchy, the overall function Fo: to transform acceleration to change in voltage is decomposed to the five following sub-functions.

Fs1: to transform acceleration to force

Fs2: to transform force to displacement

Fs3: to transform displacement to change in distance

Fs4: to transform a change in distance to change in electric current

Fs5: to transform change in tunneling current to change in voltage

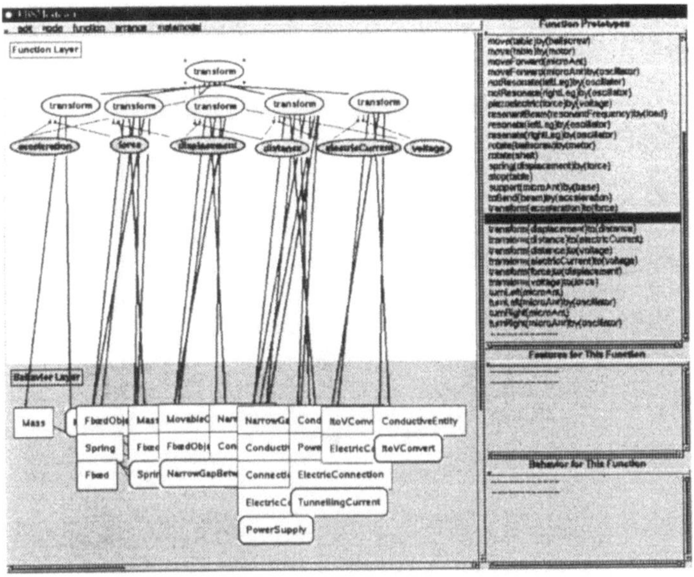

Figure 1. FBS model of the tunneling-current accelerometer

The functional components Fs1 to Fs5 are sought by SYSFUND. To commence the search for functional components, the user selects Fo in the display and selects "functional search" in the menu. SYSFUND comes up with three answers, which are the principles of the resonance accelerometer, the capacity-sensing accelerometer, and the tunneling-current accelerometer.

The last sub-function Fs5 is realized by an evaluation unit and not transformed into a physical feature in the FBS model. For other sub-functions, solution patterns are stored in the physical feature database. On the screen, when the user selects one of the sub-functions, SYSFUND shows associated physical feature(s) for instantiation. For Fs1 to Fs4, P1 to P4 are associated as realizing physical features.

P1: motion of a mass caused by a force

P2: distortion of a cantilever connected to a mass

P3: gap between a movable surface and a fixed surface

P4: tunneling electric current across a gap

The scale-analysis module is depicted in Figure 2. When the user selects a physical phenomenon in the FBS model and chooses scale analysis in the menu, this module starts with the equation associated with the selected physical feature. Values of parameters of the equation are presented in sliding gauges and numbers. One of the parameters can be chosen as the target parameter. Whenever other parameters change, the new value of the target variable is calculated and reflected in the display. Numeric computation is performed by mathematical running in the background.

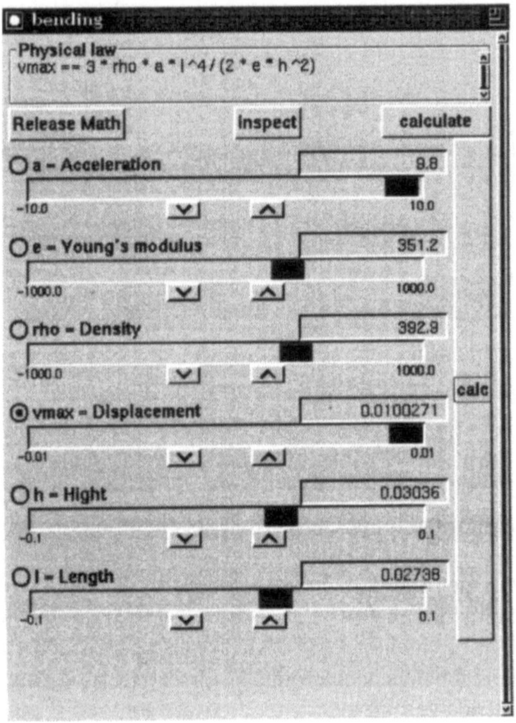

Figure 2. Scale analysis module

Process knowledge base is an important element which should be included in the concept of micromachine design system. In the case study of the design process of micromachines, we recognized that knowledge about manufacturing processes is extremely important for the design of micromachines. Without good understanding of available fabrication processes, one can not create a viable solution nor evaluate time and cost of manufacturing. It is often the case that process knowledge is dependent on the available facilities of the manufacturing site. This means that excessive abstraction to purely symbolic representation reduces the utility of process knowledge. Instead of describing knowledge solely symbolically in SYSFUND, we used HTML (HyperText Markup Language) to build a process knowledge base of micromachine fabrication called Hyper Process Knowledge Base.

The database is published on the WWW (World Wide Web)2. SYSFUND makes access to Hyper Process Knowledge Base using URLs (Uniform Resource Locators) of the descriptions of relevant fabrication processes.

Description of a Cr etching process in Hyper Process Knowledge Base is shown in Figure 3. From SYSFUND, the user may associate a structural component with URLs of relevant fabrication processes. When a particular page is requested by the user, SYSFUND uses a WWW client such as Mosaic or Netscape to fetch the data and display on the screen. Pictures can be clicked to review associated video clips that show how actual operations are done.

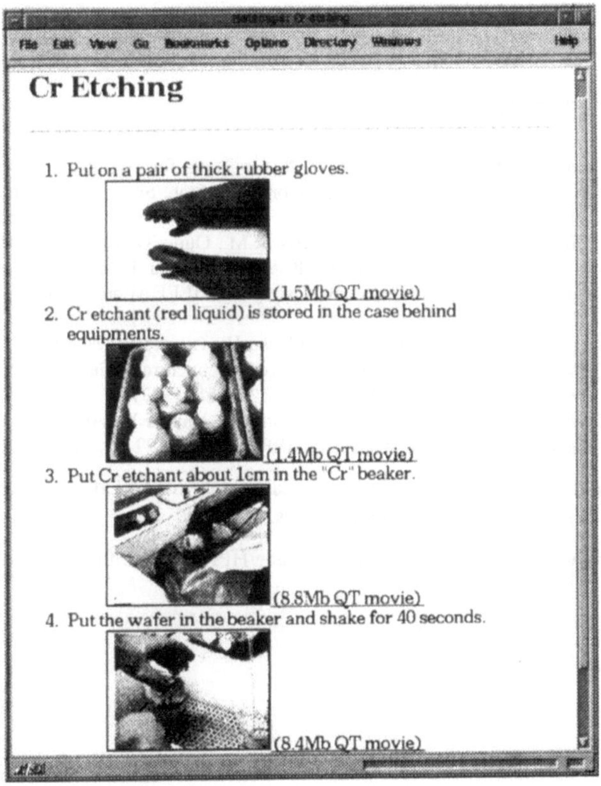

Figure 3. Hyper process knowledge base

CONCLUSION

The implementation of knowledge for the design of a micro accelerometer showed that a knowledge-based system such as SYSFUND can support the solution search. There are also many other features recognized as necessary, including the capability of analysis, material databases, and the use of process knowledge. Together with symbolic modeling and reasoning capability, these elements will be added to the concept of the integrated design environment for micromachines. Following the first experimental implementation of design knowledge in SYSFUND, the extended concept of micromachine design system is currently being elaborated and implemented at the University of Tokyo.

KEYWORDS

Micromachine design system, SYSFUND, accelerometer, HTML, etching, materials

REFERENCES

[1] Crary S., Juma O., Zhang Z., Software Tools for Designers of Sensors and Actuators CAE Systems, International Conference on Solid-State Sensors and Actuators (Transducers 91), pp 498-501 (1991)

[2] Gilbert J.R., Osterberg P.M., Harris, R.M., Ouma D.O., Cai X., Pfajfer A., White J., Senturia S.D., Implementation of a MEMCAD System for Electrostatic and Mechanical Analysis of Complex Structures from Mask Descriptions, IEEE Micro Electro Mechanical Systems, pp.207-212 (1993)

[3] Tomiyama T., Kiriyama T., Yoshikawa H., Conceptual Design of Mechanisms: A Qualitative Physics Approach, in Kusiak A. (ed.) Concurrent Engineering: Automation, Tools, and Techniques, John Wiley & Sons., pp. 131-151 (1993)

[4] Yoshikawa H., Tomiyama T., Kiriyama T.; Umeda Y., An Integrated Modelling Environment Using the Metamodel, Annals of CIRP, Vol 43/1/1994 pp 121-124 (1994)

1. Sysfund was developed at Tomiyama laboratory of the University of Tokyo. It is distributed as a free software for academic purposes.

2. http://www.race.u-tokyo.ac.jp/RACE/race/HPKB/index.html

Creation of Extended Generic Knowledge Spaces from Random and Limited Experimental Data. Application to ^{13}C NMR Structure Elucidation

Michel CARABEDIAN and Jacques-Emile DUBOIS

Institut de Topologie et de Dynamique des Systèmes, associé au CNRS, URA-34, Université PARIS 7 , 1 Rue Guy de la Brosse, 75005 Paris, France

Abstract

In structure elucidation based on ^{13}C NMR, normal empirical knowledge, deduced from reference data, outlines a very limited field of application in the space of real research. In order to overcome this severe restriction, potential knowledge is defined by extrapolation. It specifies topological neighbor relationships between primitives at the structural level, and creates a preliminary organization of the research space. Decision surfaces, consisting of spectral maps describing the behavior of primitives, are enhanced by these relationships. They simulate the physical boundaries of the maximal potential research space accessible from the knowledge extracted from the available data.

Resume

En élucidation structurale à partir de la RMN ^{13}C, les connaissances empiriques usuelles, déduites de données de référence, déterminent un champ d'application très restreint sur l'espace de recherches réel. Des connaissances potentielles sont définies par extrapolation pour lever cette restriction importante. Elles explicitent au plan structural les relations de voisinage entre primitives et organisent préalablement l'espace de recherches. Des surfaces de décision, formées des cartes spectrales décrivant le comportement des primitives, sont enrichies grâce à ces relations. Elles simulent les frontières physiques de l'espace de recherche potentiel maximal accessible à partir des connaissances tirées des données de référence disponibles.

1 Introduction

The determination of the structures of chemical compounds from their ^{13}C NMR spectra is one of the rare fields where there is a very great disproportion between the potential research space and the knowledge available to explore it. Faced with the vastness of this space, resulting from the diversity and the complexity of the open set of compounds, and the absence of real expertise based on a theoretical model, we are obliged to fall back on empirical knowledge. The first significant stage of this approach was marked by the development of spectroscopic data banks. However, the knowledge extracted from more-or-less numerous experimental data is inevitably not enough to guarantee an exhaustive exploration of the research space. Only part of this latter, centered on the reference population used and including compounds related to or similar to this population, is accessible.

In order to overcome this restriction, which is responsible for the frequent silences of a system, we have introduced the idea of potential knowledge [1]. This is not a simple transcription of the experimental facts but an extrapolation of all the facts which can be deduced from the latter. This knowledge, which is stated explicitly at the structural level in terms of generic primitives and neighbor relationships between these primitives, allows a considerable extension of the field of application of our EPIOS system. The physical organization of this knowledge on the basis of relationships established between the primitives provides an overall representation of this new research space to the system. To guide its exploration, decision surfaces, which are used to select the primitives relevant to a target structure, must also be completed. In their experimental version, these surfaces describe only the responses of the primitives to their experimental environments in the reference structures. The object of their enhancement is to describe exhaustively the behavior induced on each primitive by all its potential environments in the possible target structures. We describe here this enhancement of the decision surfaces, forming the spectral component of knowledge in the EPIOS system.

2 Structural Knowledge and Organization of Primitives

In EPIOS a structure is not perceived as a specific organization of a set of atoms by their bonds. It corresponds to a set of generic ELCOb primitives, centered on each atom one after another, and organized by their partially overlapping connections.

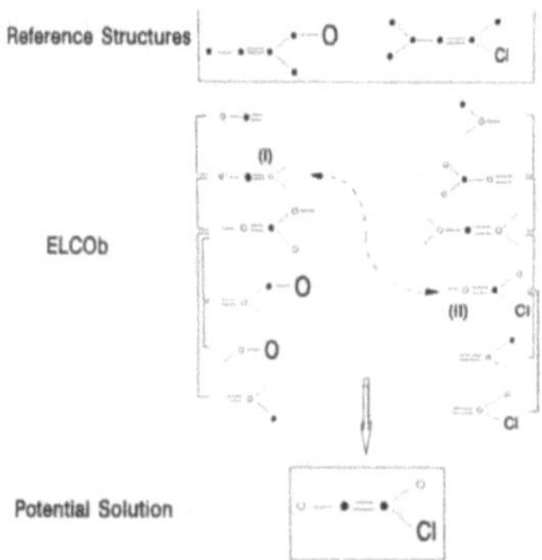

Figure 1. A new structure corresponds to new connections between its primitives.

Consequently, the elucidation of an unknown structure requires, apart from the selection of the set of necessary primitives, the establishment of all their connections. Though formed from the same set of primitives, the candidate structures proposed by EPIOS are distinguished from the reference structures by the set of new connections activated between these primitives. These experimental or potential connections were established during the learning stage of the system: 740 753 relationships were established among the 8 587 available primitives. Less than 10% (61 449) of these relationships are really observed in the reference structures. These relationships are stated explicitly in the organization of the knowledge, which provides a generic and overall representation of the research space of the system. In this way each primitive knows, a priori, by means of its latent relationships, all the neighbors with which it can appear in a possible target structure.

To scan the research space represented by the primitives and these relationships, spectral maps, characteristic of the effects of environments on the primitives, are used as decision surfaces. They control the selection of primitives during the interpretation of a query spectrum. Their progressive assembly towards the candidate structures is then guided by the activation of their relationships with the neighborhood. They are initially established on the basis of the spectra of reference structures, and describe only the environments corresponding to relationships which are really observed (10%) in these structures. The aim of enhancing them is to describe all the environments likely to be formed from an exhaustive set of potential relationships.

3 Definition of Extended ELCOc Environments

Formally, a chemical structure can be considered as a concentric expansion around any one of its primitives, taken as the origin. Ideally, the complete set of possible structures starting from this primitive would be necessary to establish its entire spectral map. This set is open and cannot be fully itemized. On the other hand, the elementary expansion of a ELCOb primitive leading to extended ELCOc environments (Fig. 2) makes it possible to give an exhaustive account of the structural contexts possible in this set of structures.

Figure 2 illustrates this concentric growth leading from the ELCOb primitives to the structures in which they are likely to appear.

The ELCOc environments represent the first step in the generation of any structure from one of its ELCOb primitives. In practice, they are obtained by combination of its potential neighbors around each ELCOb. These combinations put into concrete form the 740 753 neighbor relationships established previously between the 8 587 ELCOb primitives available. For each one of them the number of possible ELCOcs is the product of the number of relationships associated with the neighboring atoms. With all the potential relationships, 246 923 053 ELCOc environments are generated whereas only 49 193 environments of this type appear in the reference structures. This difference gives an idea of that between the experimental knowledge and a generic representation of the potential research space of an elucidation system.

However, not all the ELCOc environments created correspond to chemically stable entities. Nevertheless, for the system, which is ignorant of the idea of chemical stability, they constitute so many possible environments which can be taken into consideration during an elucidation. In order to make these ELCOc environments accessible, and beyond the candidate structures which they can form, the spectral maps characterizing their influence on the ELCObs are now updated.

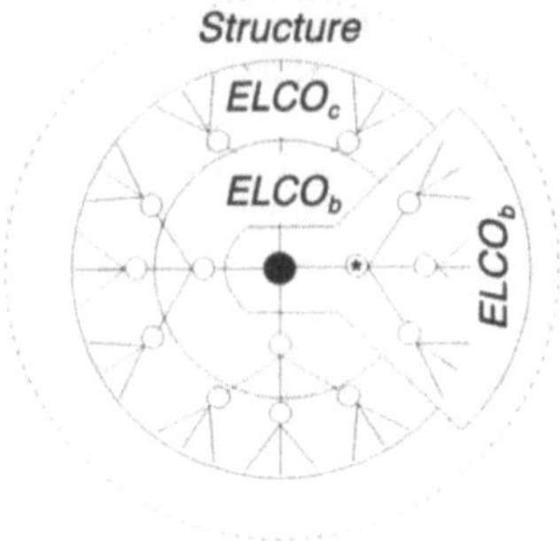

Figure 2. Concentric expansion of ELCOb primitives towards target structures.

4 VIRTUAL SPECTRAL MAPS

The relationships used to construct ELCOc environments about the ELCOb primitives are used to describe their influence on the behavior of the latter. Through these relationships, each ELCOb primitive, formed by N pairs of atoms (Fo,Ai) is associated with N sets of potential neighbor ELCOb primitives centered on the Ai atoms. This simultaneous presence of atom pairs, in transposed forms, in an ELCOb and in its neighbors, is exploited in order to define the influence of the ELCOc environments. The behavior of an ELCOb is then not described completely by its response to its experimental environment. It corresponds also to the behavior which its presence could induce in its neighbors.

Figure 3 represents the practical results of this method applied to the ELCOb formed from two pairs, cc_1 ($= o - o \leq$) and cc_2 ($- o = o -$). The map in Figure 3-a brings together 350 pairs of experimental values (δFo, δA1) assigned to this ELCOb in the spectra of 261 reference structures. In these structures, the ELCOb considered is observed in only 71 experimental ELCOc environments, whereas 1 064 potential ELCOc environments are associated with it. To characterize the latter, the previous spectral map is completed by considering the behaviors induced by all the neighboring ELCObs, associated via its couple, cc1(Fig. 3-b). The second couple, cc2, is treated in exactly the same way.

A total of 502 and 1 782 pairs of (δFo, δA1,2) values are brought together to establish the maps describing the behavior of the considered ELCOb in its potential environments.

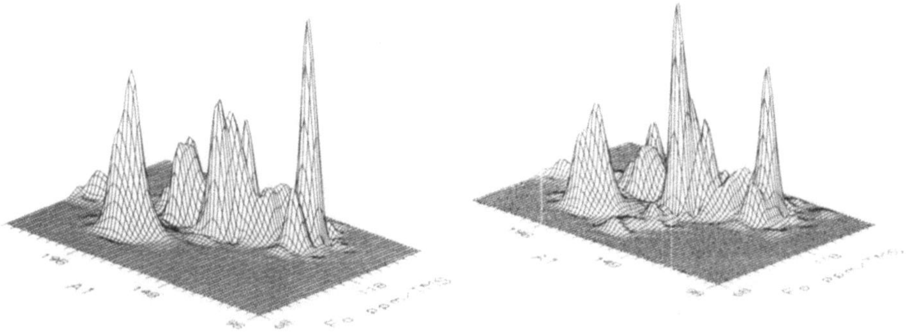

Figure 3. (Fo,A1) maps: (a) experimental (b) potential.

Apart from the modification of the contour of the experimental maps, the effect of the potential environments is expressed, in the example proposed, by the appearance of a new region of behavior for this primitive. The behavior defined by this region, δ 13CFo $\in$ [90-110], δ13CA1 $\in$ [160-175], (Fig. 3-b) is the response of the given ELCOb to the effects of potential environments which are absent from our reference population, of the type:

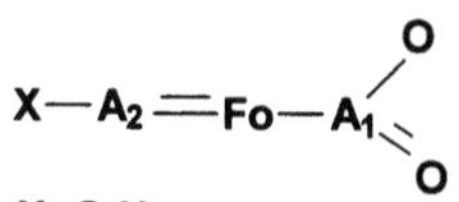

X=O,N

Although it is approximate, this behavior deduced from sets of very varied values corresponds with sufficient accuracy to that predicted for this type of environment.

CONCLUSION

The overall enhancement of knowledge resulting from the establishment of potential spectral maps is illustrated by the comparison given in Figure 4. It represents on the correlation planes the number of ELCOb primitives experimentally (Fig. 4-a) and potentially (Fig. 4-b) associated with each of the pairs in these planes. For the 14 663 pairs of carbons, (Fo, Ai), forming the 8587 available ELCOb, 1 030 833 (δFo, δA1) pairs are used to establish their potential spectral maps. The increase in this number expresses the extension of the spectral maps by the new behaviors attributed to the ELCObs as probable responses to the effects of their potential environments. This extension of the maps determines that of the solution space of the system.

The exploration of this extended space involves the treatment of an increased number of alternatives during the search for candidate structures. Coping with the noise they generate is the price which must be paid for the advantages related to the minimization of the silences inherent in the use of purely experimental knowledge.

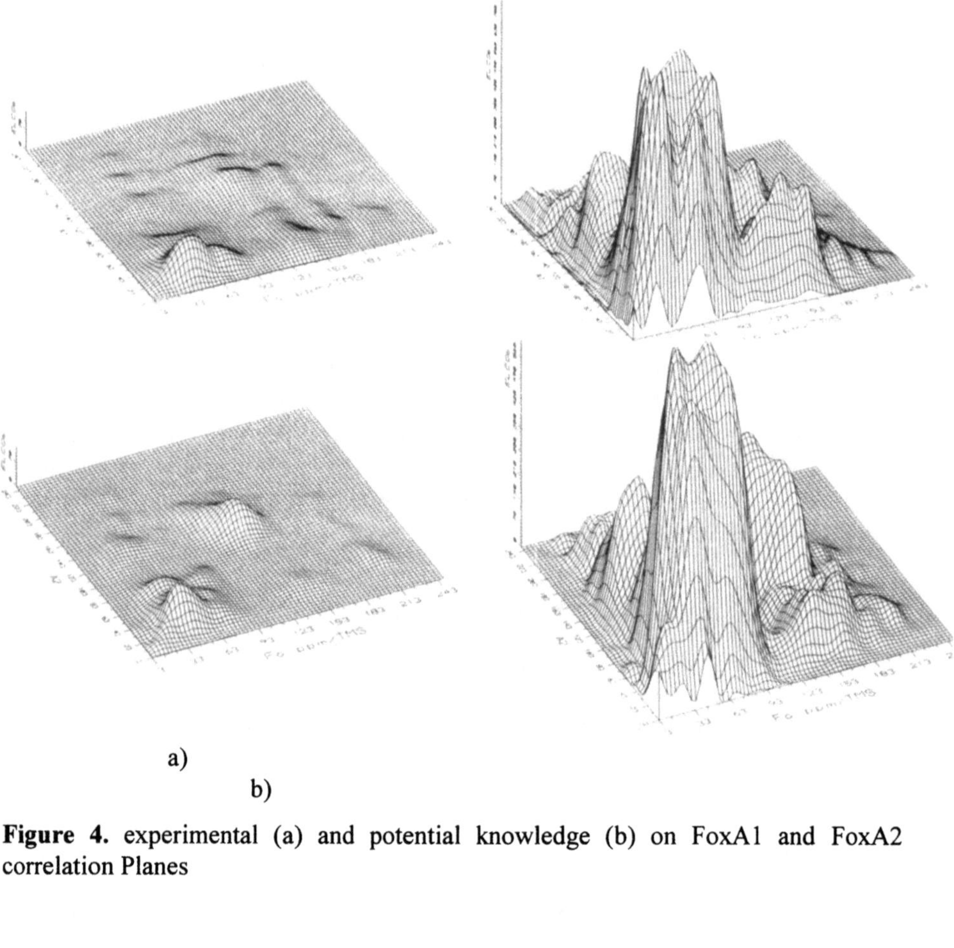

a)

b)

Figure 4. experimental (a) and potential knowledge (b) on FoxA1 and FoxA2 correlation Planes

KEYWORDS

Structural elucidation, virtual knowledge, spectral map

REFERENCE

[1] Carabédian M., Dubois J.-E., Inferring Extended Virtual Knowledge from an EPIOS Conversion Graph of Overlapping Substructures, J. Chem. Inf. Comput. Sci., 14, 701-706 (1994)

INTEGRATION OF MATERIALS DATA SYSTEMS, FUSION OF KNOWLEDGE AND DATA, AND DYNAMICS IN MATERIALS DESIGN

Shuichi IWATA

Research into Artifacts, Center for Engineering, The University of Tokyo, Komaba 4-6-1, Meguro-ku, Tokyo, 153, Japan

ABSTRACT

Materials design is discussed as to its procedure and implementation to discover hidden potentialities of materials developments on the basis of huge and high quality data bases and effective knowledge. Rules on materials properties are categorized by structural primitives and model primitives. Procedures for presuming, reasoning and verification of materials design are discussed to find an exemplar of strategic integration of materials data systems, where fragmented information of different objectives is to be "melted" into new artifacts of new values. Systematization to increase freedom in such procedures is to be realized as dynamics in distributed computational environments operated by real experts.

RESUME

La conception de matériaux est discutée aux niveaux des procédures et de leur implémentation pour découvrir les potentiels des développements de matériaux à partir de grandes bases de données de haute qualité ayant une connaissance effective. Les règles sur les propriétés des matériaux sont classées par des primitives structurales et de modèles. Diverses procédures pour prédire, interpréter et vérifier la conception des matériaux sont discutées pour dégager un type d'intégration stratégique de systèmes de données de matériaux dans lequel les informations fragmentées provenant d'objectifs divers seront "fusionnées" dans des artefacts originaux. La recherche d'une systématique génératrice de degrés de liberté pour aborder les procédures évoquées doit être menée de façon dynamique par des experts chevronnés dans un environnement informatique distribué.

INTRODUCTION

In spite of the huge amount of information available in different forms, everyday we face new discoveries, unexpected accidents, mistakes, knowledge guessing;

enlightenment procedures in designing materials essentially imply a broad perspective for the universality of presuming, reasoning and consequent problem solving. There are some similarities for the following problems.

- Designing new materials and artifacts based on fragmented information and tools,
- Decision making in managing enterprises under uncertainties of the market,
- Trouble shooting against failures and accidents in a complex system,
- Watching for undetectable enemies,
- Recent environmental issues of ambiguous causalities,
- Preventive medicine for newer diseases.

Our goal now is to establish a method of designing suitable materials for any given requirement. To get an ad hoc solution for each problem is relatively easier than to solve it systematically. If we can succeed in solving the typical complex problems in materials in an explicit way, this might have a big impact on other fields and such is the target of this research.

To solve such problems, there are several reasoning methods and computational environments to discover tactics and strategies, e.g. learning and heuristics, generic algorithms, neural networks, rough sets, the Dempster-Shafer theory, qualitative reasoning, naive physics, knowledge acquisition by regularities by clustering/classification/ articulation, Cybernetics, abduction machine. Exemplars have been reviewed in many articles, e.g. by N.R.Hanson[1] and by N.Wiener[2].

Some mathematical approaches may exist as a "detective" story for solutions as implied in M.F. Ashby's approach to systematizing procedures of materials selection from maps on complexities of material performances[3] and in P.Villars's approach to discovering regularities in a stability of intermetallic compounds[4],[5].

Main points to be dealt with carefully exist in procedures on :

- How to reduce the dimension,
- How to derive a new dimension,
- How to deal with different representations of the same semantics and/or the same representation of different semantics.

In this paper a new possibility for discovering hidden potentialities of materials developments from huge data is proposed, reflecting the author's understanding on recent-namely '90s-drastic developments in data bases, first principles calculation, distributed computing, computer networks, etc.

2 INTEGRATION OF MATERIALS DATA SYSTEMS

Complexities of materials are usually difficult to describe with full physics, and practical approximations are to be introduced as shown in the schema. They are described, empirically and/or with certain physical frameworks depending on our understanding, in spatial variation of stress, temperature, or concentration, which may require discrete part computations-articulated with respect to spatial structure- for their solution. Different levels of articulations have been introduced. On structures, we can list primitives going from microscopic to macroscopic.

Structural Primitives
{electron/nucleus,
electron gas/lattice,
atom cluster,
unit cell/crystal,
{lattice/defects
-point defects,
-line defects,
-boundary,
-surface,
-volume defects},
defects/defects,
grain/grain boundary,
matrix/precipitates,
matrix/particles,
matrix/fibers,
parts,
components,
system,
plant,
environment}

Almost all structural primitives are in principle listed as a combination of some primitives and can be estimated by using some features by first principles calculation. Recent data bases also cover their general features. The main concern here is to pick up the essential ones, especially in mesoscopic domains as better compromises between microscopic and macroscopic views.

Views for dealing with data sets, for example, are to be categorized in the following way.

Views
{static/dynamic,
deterministic/probabilistic,
microscopic/macroscopic,
instantaneous/hereditary,
Eulerian/Lagrangian,
precision}

Time dependent phenomena are usually the main concerns in practical applications, e.g. fatigue, creep, corrosion, reliability, life extension. Due to the huge diversities of materials performances, strategic problem solving has been carried out and some parts are prepared as "manuals" and "expert systems" through extracting exemplars from well-known and well-defined experiences.

Relations between structural primitives can be described by geometrical meanings and chemical semantics.

Relations
{{configuration/structure
-coordination,
-arrangement,
-orientation,
-pattern},
{interaction
-binary,
-ternary,
-multiple},
correlation,
causality,
total}

Theoretical approaches are usually summarized from the ideal and simple case, and they usually do not reach complexity as a total description. Qualitative reasoning based on theoretical frame works and quantitative estimation based on empirical fitting are usually a practical compromise utilizing available information. Identification of states of materials is carried out by declaring temperature, pressure, compositions and so on.

For classification of materials information, this kind of information is utilized if we can succeed in identifying each state explicitly, e.g. temperature and stress distributions of a turbine blade and consequent microstructural evolution and deformation are key information for the life evaluation of the turbine.

Properties of a material can be described as dependent variables of the above definitions, e.g.,

Property
{mass,
shape/size,
density,
crystal structure,
electronic,
electric,
diffusion,
thermal,
magnetic,
{mechanical

-static strength,
-toughness,
-stiffness,
-hardness,
-elastic,
-creep,
-fatigue,
-dumping capability,
-super plasticity,
-friction and wear},
optical,
fabricability,
chemical,
corrosion,
processing,
shape memory,
biological compatibility,
availability}

= f
(Structural Primitives{.},
Views{.},
States/Environment{.},
......)

For example, in the case of composite materials, parts and their boundaries are introduced and arrangements of such parts are optimized to attain required performances.

f
(Structural Primitives {matrix(Al)/fiber(SiC)},
Views{precision(order estimation)},
Relations{bulk},
States{temperature(200C), rocket body})
= Properties{static strength(1500 MPa)}

Let us assume a causality in a certain data base between the number and density of fibers and static strength. How can we find such a causality (=how can we reduce the dimension), how can we find new parameters to improve properties (=how derive new dimensions, e.g. lattice mismatch between fiber and matrix at the boundary), how find new patterns to change the strategy of materials developments (=how deal with different representations of the same semantics and/or the same representation of different semantics, e.g. using the pattern of fibers in bamboo for the design of fiber arrangements)-from available data bases to improve materials properties with confidence? Different aspects are effectively evaluated and utilized to solve the real problem.

The following examples will imply another aspect of modeling. Complexities of fluid flow are reproduced by a lattice gas automata approximation as another representation of Narvier-Stokes equations and molecular methods like direct Monte Carlo simulation were recently applied to explain critical hydrodynamic phenomena.

The finite element method is another example of articulation for continuous bodies. The discrete parts respond to temperature, body forces, and stress via constitutive equations, with the constraints of equilibrium, compatibility, and continuity imposed at their boundaries. Elegant assumptions are prepared to solve engineering problems, practically solving areas where continuous models are valid without solving the tough problems in the vicinity of singular points, namely, crack tips where atomistic and electronic considerations are to be taken into account.

If the level of articulation is divided down to the atomistic level, it becomes the molecular dynamic calculation where the atom-atom interactions are described in terms of potentials. Energy dissipation through Newton's law is calculated effectively, but we need corrections on the electronic energy loss. Its strict calculation is left to first principles calculations.

Structural stability of materials parts and their dynamics are simulated with flexible resolutions as described above but with limitations on time and size. One practical important fact inferred from these examples is that there are many apparent complexities that can be described by combinations of simple and well-known available models.

It should be kept in mind at each step that there are two kinds of models, namely, empirical fitting models and theoretical models. The articulation process on material complexities with respect to underlying mechanisms is very sophisticated but is usually carried out as a resolution of a problem into solvable sub-problems, namely, pattern matching along structural resolutions to obtain more comprehensive solutions. Phenomenological correlation after identifying the problem is carried out as the first approximation, and afterwards sensitive variables are extracted through intelligent simplification without losing certain essentials that really matter.

There are many models on materials properties/performances and corresponding computer programs, and some of them are regarded as exemplars and allocated to solve suitable sub-problems.

If we gaze carefully at a set of complexity data, sometimes we can find some essential similarities between the selected data set and the given problem. In such a case we use it as a reference. If a model is only a concise description of a body of experimental data, we must be careful as to its valid domain. If it captures the essential physics for the given problem, the model can be used as a tentative substitute for experiments. If we can resolve the complexity into essential parts and relations and if we can make a transfer model from which the reality of the given problem is reproduced by the resolved parts and relations, we can reach an entry point to solve the problem of complexity.

3 FUSION OF DATA AND KNOWLEDGE

Arrangements of model and structural primitives to elicit a suitable or possible strategy for solving the given problem are carried out in the following way.

1. Formation of a Skeleton Line

The first phase of integrating available information involves regrouping the information by suitable parameters that relate to material properties-for example, by texture, to describe anisotropic mechanical properties, and by charge transfer, for stable surface structures. After compiling the information into appropriate groups, a set of rules emerges to form an information base consisting of knowledge and data to describe materials properties. A skeleton of the virtual production line is formed by rearranging available information in the information base (data bases and knowledge bases) with respect to controllable parameters and results. The complexity of materials allows only an imperfect skeleton of the line to form in the first phase, which uses materials data and assumptions adequate for a first approximation. In the second phase, this skeleton is refined by semi-empirical rules and suitable theories.

Next, a stepwise procedure is proposed to organize the materials data for a materials design which integrates all relevant information-namely, data and models-according to:

(1) causal relations among raw materials, processing, and resulting microstructures and properties, (2) value-added cycles from raw data, compiled data, evaluated data, design data, and performance data, and (3) resolution of a macroscopic continuous view to substructures consisting of microstructures, defects, and finally atomistic levels. Sometimes subsets of data are replaced by ad hoc models of equivalent meaning.

Ideally, materials data are characterized at a certain comprehensive level, for each application which is described, by a set of relationships between relevant data groups. Typical relationships are grouped into the following pairs :

(Objectives, Performance)
(Performance, Properties)
(Properties, Macro- and Microstructures)
(Structures, Processing)

Materials design travels a downward path from "objectives" to "processing", while the causal relation goes upward from "processing" to "objectives". The former corresponds to design processes that seek suitable methods to achieve goals, meet specifications, or determine causes of practical problems. This could be called an abduction line or a resolution line depending on the description of the objective.

A production line is based on actions and results. In an abduction line, the objective set is a target of materials developments, and the process is to find a solution. In a resolution line, the objectives are described as materials troubles to be analyzed. For example, the causal relations on the effects of titanium added to AISI 316 stainless steel are described as follows :

Rule 1: If Ti is added to AISI 316 steel, compressive stress states occur in the vicinity of titanium carbide precipitates.
Rule 2: If the titanium-modified AISI 316 steel is manufactured by Process A, then we get fine titanium carbide particles effectively dispersed in the matrix.
Rule 3: If there are compressive stress states, then the flow of vacancies and interstititals is biased.
Rule 4: If vacancies are dispersed in the matrix, then swelling due to vacancy aggregation is suppressed.

By combining Rules 1 to 4 in different ways, we get the skeletons of production, abduction and resolution lines.

Production Line: Titanium addition -> Process A->Swelling reduction
Abduction Line: Swelling reduction-> Titanium addition -> Process A
Resolution Line: Swelling -> Aggregation of vacancies -> Biased flow of vacancies

The idea of a "virtual" production line was proposed as a collection of causal relations screened by objectives. A skeleton of the production line is prepared as a qualitative/conceptual description of a causality sequence of three terms, namely, input, mapping and output. Starting "primitives" and "a product" are extremes of the causality sequence. First, relevant retrieved data are rearranged along typical causalities. Then the line is "turned" in the refinement phase, as described next.

The structure is analyzed on various scales, depending on properties, performance, and materials, e.g. homogeneous objects with cracks, composite materials with textures, polycrystals with fine grains, single crystals with high dislocation density, valence electron/atom, electron=state density, or isotope ratio.

After the preceding preparation, ad hoc linkage of available information into a production line becomes the objective of the system building. The purpose of this prototype data system is to find the most "suitable line" for precisely known information and to prepare a "tentative line" for areas still uncertain but awaiting solution. It is essential for the production line to have flexibility with respect to (1) combinations of elements and sub-components, (2) types of microstructure, (3) processing methods, and (4) combination of properties.

Predictions of the synergistic effects of the "produced artifacts", after combining starting sub-components, are left as the key issues for further refinement. As a tentative solution for setting up this production line, a database with a data model connecting items along the causal relations is used, where binary relations such as :

processing -> microstructure
microstructure -> property

are intentionally extracted from the database,

2. Refinement of the Skeleton Line

To refine the production line as to efficiency and quality and to go from qualitative predictions to quantitative descriptions using dominant parameters, four issues need to be addressed. These issues are key steps and their solutions are proposed from a practical point of view based on state-of-the-art databases and knowledge bases

Usually the skeleton of the production line is not perfect and does not result in the materials that are sought. This is because insufficient or contradictory information causes discontinuity in the skeleton line at several points. These discontinuities can be bridged by complementing available data with crude induction based on phenomenological correlation among available parameters. The procedure for refining the qualitative skeleton is as follows :

(1) Finding causal relations implied by a set of relevant data. In order to formulate a continuous production line targeted for use, relevant information must be exhaustively retrieved by taking advantage of set/subset relations between different objects e;g.: hcp metals/zirconium, titanium; interstitial in refractory metals/impurities in molybdenum, vanadium, and so on. These relations are defined as metadata for effective use of available data.

(2) Verification and refinement of the selected data, based on common knowledge. Observation and classification of the selected data are at first carried out by selecting a set of well-characterized data introduced on the basis of common knowledge at each design window. An example is the optimization of the biased flow of vacancies and interstitial in solids to reduce swelling. To retain "common sense" about materials and to ensure that no possible materials are overlooked, overviews of generic data-the big picture-are compiled at the beginning of refinement. Different overviews with different windows give us a good chance to browse through available information.

(3) Resolution into parts and analysis of the selected data. Underlying mechanisms are sometimes explained in terms of the substructures, which are used for modeling, grouping data, and starting "primitives" for further reconstruction. Relations such as "Part of" the dependence among variables (e.g., the titanium carbide "parts" of stainless steel and the biased flow of point defects) are used to obtain more comprehensive explanations of the production line, based on available information.

(4) Selection of methods to reconstruct artifacts from predefined parts. After selecting primitives and the methods for reconstructing such primitives into artifacts, the outcome is evaluated to determine the capabilities of predicting reconstruction effects. To achieve a good compromise between steps (3) and (4), namely, to select suitable starting primitives for practical models, is the focal point of refining the production line as a feasible procedure. Results of the preceding refinements are to be completely evaluated by examining the properties of produced artifacts and the efficiency of the production line.

4 DYNAMICS IN MATERIALS DESIGN

Almost everything described above is a working guideline to be verified by "doing by learning", which cannot be solved by one powerful method and requires a dynamic system utilizing different relevant resource-data bases, meta data, knowledge bases, computational codes, objects-effectively. Effective procedures in general are divided into the following four steps,

(1). Get-the-Big-Picture on the given problem so as not to miss undiscovered potentialities by taking advantage of huge distributed data systems.
(2). Get-the-Precise-Picture on the materials so as not to miss in-situ feeling on materials, e.g. by a so-called virtual laboratory of flexible resolutions from macroscopic to microscopic aspects integrating available data and knowledge,
(3). Get-the-Whole-Picture within the domain of the given problem through complementing imperfect information by suitable working hypotheses, and
(4). Get-the-Practical-Picture based on needs and design specifications.

Step (3):Get-the-Whole-Picture consists of interpolation and extrapolation. The interpolation is carried out by suitable grouping/classification and the extrapolation by effective trial-and-error/ heuristics/ fluctuations/ break-through/ inspiration/ intuition/ something. The latter is to be realized by an abductive environment of the highest flexibility. The fourth step is to deal with the problem of value which might be very difficult to share but should be shared at a certain level for each case. Thus, the design environment is realized as a "melting pot" where each unit of information sources is linked dynamically, preferably by computer networks, to solve given problems. For this sort of linkage, standards for communication, data description and knowledge representation are essential and strategic guidelines as proposed in this paper are one option for cost-effectiveness.

KEYWORDS

System integration, materials data base, materials model, materials design, materials data system, virtual production line

REFERENCES

[1] Hanson N.R. , Patterns of Discovery, Cambridge Univ.Press (1958)
[2] Wiener N. , Inventions, The MIT Press (1993)
[3] Ashby M.F. , Materials Selection in Mechanical Design, Pergamon Press (1992)
[4] Villars P. , J.Less-Common Met. 119, 175 (1986)
[5] Villars P. , The Structures of Binary Compounds edited by F.R. de Boer and D.G.Pettifor, Elsevier Sci. Pub., 1-103 (198?)

Chapter 2

ARTIFICIAL INTELLIGENCE: IN KNOWLEDGE-BASED SYSTEMS

HUMAN DIMENSION IN ARTIFICIAL INTELLIGENCE. TASTE, PREFERENCE, CHOICE - BEYOND CLASSICAL OPTIMIZATION

***Vincent* DOUZAL**

Cemagref, BP 121, 92185 Antony Cedex, France

ABSTRACT

When working to bring a recipe to perfection, a cook is clearly not in a world as simple as that of arithmetic, where 2+2 always equals 4. How can a scientific method help him to formulate a new recipe, that is to guess what will be preferred by our sensory system? A usable representation of sensations implies building a specific language permitting a very restricted optimization.

RESUME

Quand on cherche à parfaire une recette de cuisine, on n'est pas dans un monde aussi simple que celui de l'arithmétique, où 2+2, quoi qu'on fasse, égale 4. Quel peut être l'apport de méthodes scientifiques au jeu de trouver la préparation qui sera préférée au travers de nos sensations gustatives ? Une représentation correcte et manipulable des sensations passe par la construction d'un langage, et permet d'optimiser dans un sens très atténué.

1 INTRODUCTION

The present article is on the whole concerned with extremely common activities: cooking, and eating, preparing a recipe, then tasting. These practices are usually thought of as an antinomy to science. French for instance is fraught with expressions that oppose informal cookery to the systematic march of science. To what extent can designing a recipe be subjected to a scientific process? This is the question we want to address.

In the first part, we try to describe, from a very general viewpoint, some aspects of cooking and ways to formalize them. In the (nearly unavoidable) case of seeking the "best" taste, we are led to define in what sense cookery is subject to optimization. Therefore, we must first find how an edible sample can be correctly identified, characterized; we must be able to run comparisons and ratings. This is only possible through a description of a taster's feelings. The second part explains in what a disappointingly restrictive sense this can be done. A third part exemplifies those principles on an automated apparatus-aided formulation project (initially and pompously called "the artificial cook"). The last section then discusses the meaning of optimizing a recipe —or any sensorily-characterized product—, and shows how different it is from the usual framework of optimization.

2 WHAT DOES FORMULATING, PREPARING, A RECIPE CONSIST IN?

In Western civilization, the food industry is challenged by the modern trend for newer things, and has to face an unprecedented pursuit of novelty. The evolution of customs, techniques for goods production, food processing, conservation, regulations on public health, all share in an inextinguishable desire for new products.

The competition between companies spans over advertising, brands, etc. But at the very core, a product's capacity for success, its market development —or stability despite this permanent invasion— is linked to its quality: how essential or necessary it is (nutritional value), how much it is preferred due to its organoleptic properties. This last point is our concern here.

2.1 Creating and testing recipes

Inventing new recipes, however, is neither exclusive to the food industry, nor a recent concern. The discovery of fire must have been the first trigger to exploration of cooking possibilities. Since then, generations of cooks have given birth to thousands of recipes, through empirical, or pragmatic principles.

Basically, the method is trial and error. Formulation includes both preparing a dish *and* tasting it. This simplistic process has nevertheless led to the incredible diversity of ways for satisfying our taste we can find all over the world. Referring to these brilliant creations, one is generally tempted to consider cookery as an art.

There is no doubt that an artistic aspect will always be part of food preparation, but we can suppose there is room for improvement regarding scientific methods. We can identify three important trends in the elaboration of a recipe.

• *Creating something new*. This relies on the cook's ability to foresee or to translate a fantasy into edible reality. In many cases, chance unveils new possibilities. Corn flakes are said to have been discovered because of an oven left on and forgotten.

• Once an area is identified, the precise process is *adjusted*. By variations on the main theme, mastery is sought, evolving in search of *greater satisfaction*, trying to guarantee success by appropriate tricks. There are as a rule many roads to failure and but a few to success. Pleasure is known to be built, by reinforcement or inhibition. This leads to a cycle between eating and cookery customs, explaining the birth and differentiation of local tastes, much like Darwin's evolution of the species. Cooking and taste together form a historical process, capitalizing on experience and habits over the centuries by day-to-day rituals.

• The third stage is *imitating*. Because one never has exactly the same ingredients, each time one wants to reproduce one's own recipe, one imitates oneself, trying to compensate for these perturbations. To include one aspect of a dish in another recipe, one does likewise, trying to guess which combination of actions and components will operate correctly.

This surprisingly very important feature poses the problem of defining when two samples can be considered identical (how much one imitates the other).

2.2 Striving for the preferred recipe through our sensory system

We can envision modeling the cook's reasoning, finding back-and-forth relationships between preparation and perceptions. Descriptors here are necessarily in French (see Part 4 for justification, Figure 4 for translation).

The cycle —preparing dishes, tasting, choosing the most pleasant one, preparing new ones, etc.,— to seek the "best" recipe is directed by *pleasure*. By simplification, it can be expressed in terms of *preference*: In given conditions, what you opt for should be the most pleasurable.

Classically, formulation is understood as exploring and relating the spaces of compositions and preferences (Fig. 1). We understand, however, that for an ideal formulation, we need access to the taster's sensations.

The very fact of calling dishes identical or different supposes the ability to qualify (or quantify parameters after analysis) correctly how each one differs from its variants. To do so, the definition of components and the sequence of operations to which they are submitted cannot provide an appropriate description. In many cases slight differences in the recipe lead to obviously changed results. This refers to our direct experience, a relevant characterization of a product can be made only through its sensory properties. It is also an a priori idea: if an ideal thing exists for a person, it must be expressible in the words of this person's perceptions.

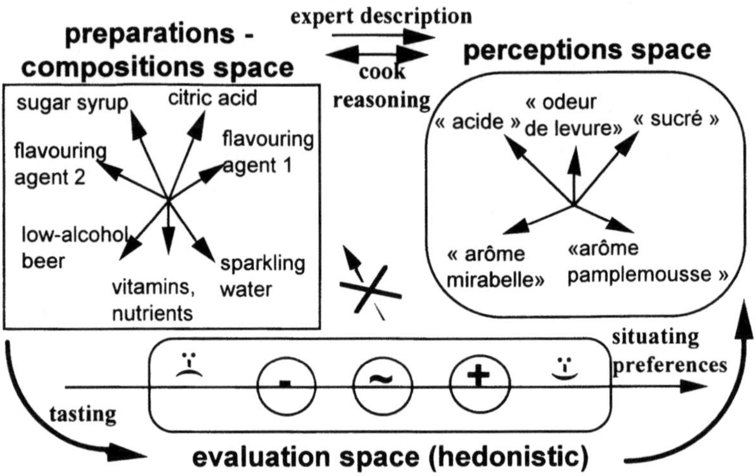

Figure 1. Formulating is interestingly represented by spaces and their relationship.

While there are numerous ways, e.g. chemical analysis or physical measurement, of qualifying food, no one of them, no measuring device, and not even all of them together, is either accurate enough or sufficient to reliably express what happens in a person's mouth, and then what sensations it arouses in his mind. Therefore, there is no way to eliminate the taster in the formulation-test loop. We need him as a measuring device. We shall evoke below how reliably a person or a group of persons can be considered as a measuring device.

Reviewing again the cooking-tasting cycle, we discover we have neglected one essential phase: The taster does not only show his preference. He talks to the cook who tries to determine in his own terms which sensations are desired, which are repulsive. These sensory descriptions are of two kinds. Some are expressions of pleasure or preference (hedonistic). Others are simple comments on sensations, developed in multiple terms and evocations, representing as many dimensions of the appearance of the food to the taster.

The cook uses the latter, seeking as he listens for a translation into ingredients and actions. Looking back at Figure 1, it is obviously impossible to guess from the one-dimensional hedonistic space where to go in the n-dimensional preparation space. It is possible to improve the recipe only by playing on the relations between the three spaces of Figure 1. The p-dimensional perception space can help assess desirable areas of the preparation space. The cook needs to calculate a partial inversion of the tasting operation, to go from desired perceptions to a recipe. This is what we want to model.

3 The Quest for the Taster's Sensations and Feelings

It is epitomized in two questions.

3.1 How can we represent sensations?

In the broadest sense, this asks what is a representation of the human mind. It has been widely and controversially debated in philosophy, then in science, and has led to the emerging sciences of cognition which try to have the many different views of the mind converging into a unique paradigm. It is not just by chance that the strongest agent in cognition is artificial intelligence; it holds sway over that modern and totemic symbol of the mind —the computer.

The most general *term* to designate an object of the mind is a symbol. Accordingly, sensations should be represented as symbols. But symbols actually seem so ubiquitous, so universal that it barely helps us.

From this abstract point of view, the most relevant discipline is probably semiology. Most of its attempts to classify all the possible sign codes involve their level of articulation, i.e. starting from the elementary signs, their capacity to combine into elaborate structures. When it comes to using a system of symbols, we have a strong and irrepressible tendency to resort to highly composable ones.

Computers are well-known to process "symbols". And the general assumption that our world is actually made of numbers, or numeric quantities theoretically allows for coding anything into a form suitable to computing.

While laying the foundations of computing theory, Turing made a decisive stride in a still ongoing revolution of the science of the mind. Artificial intelligence echoes his claim that thoughts can be mapped onto symbols. Any physical object could stand for a symbol, and it would be the same to manipulate these objects or thoughts. A computer, processing symbols, should therefore be granted a thinking capability, and even consciousness.

According to this optimistic perspective, the computer is the right tool. But we still do not know how to feed it with the information relevant to our question:

How can we access sensations? How can we access feelings?

3.2 Analyzing the physiology of perception

It is the first and natural-seeming idea that springs to mind. Taste at large consists of gustation, olfaction and retro-olfaction (smell, aroma), tactile sense (mainly on the tongue), thermic sensations (chiefly inside the mouth). All these components can potentially play a major role in the overall perception. Everybody knows how different pasta types of the same material taste different. No one would mistake sweet cream for whipped cream although the only difference is that the latter has been whipped, implying a mere texture modification.

Though chemical sensitivity is at its core, taste extends well beyond this. In fact, any chemical reactivity may well be turned into a form of chemical sensitivity of which every cell is capable. We must simplify this inquiry by focusing on the specialized cells of the well-identified sensory tract. Starting from the olfactory epithelium, going to the olfactory bulb, information flows through the thalamus, is subject to conscious analysis in the olfactory cortex (which allows us to name it), but passes also to the hypothalamus where it develops its emotional aspect, triggering all kinds of reactions e.g. salivation, fear, flight,... that pervade our behavior, often involuntarily. Odors and tastes are often directly, profoundly linked to vital messages (bitter flavor evokes a poisonous menace, which is actually well-founded, considering the properties of alcaloids found in many plants; pheromons are molecules liberated into the air that transmit messages essential to mating for animals).

There is a permanent and unavoidable drift from sensing to feeling, from a conscious analysis or description of olfactory information (allowing at least some labeling of odors and flavors) to a mainly unconscious hedonistic reaction that dominates inclinations and preferences. The latter certainly differs from one person to another; each has his own history, as is the case of the former. An important fact confirmed by neurophysiology is the existence of specific anosmia, i.e. total incapacity to smell a molecule for lack of compatible receptors. It is the strongest and clearest hint that there is no standard observer in chemosensitivity.

Contrary to the other senses (the most typical being vision), there is no common system of reference (in the physical sense), no shared "objective" topology to which we can refer our experiences. A symptom, or a consequence is our inability to designate flavors and odors; we circumvent this by naming them after their source ("it smells like mint"). We do not know any odor descriptors.

Moreover, the only objective references we can have, molecularly pure substances at known concentrations, do not help: We do not really know how to map in a topologically consistent way all known molecules; each one merits a specific dimension on its own, and we are left with no structure, no landmarks, in a gigantic space.

Olfaction gives very little room for abstraction. Naming refers to an object, and in the process, simplifies it, making it amenable to manipulation and classification. Classifying might almost be seen as the central activity of the human mind. If each odor needs its own name, if no rule can be found to designate by the same descriptors different objects, there is no way to elaborate hierarchies, to build abstractions. Odors refuse mental combinations.

Finally, the hope of obtaining a good image of sensations by neurobiology is also vain. First, many stages in creating the nervous olfactory message are still poorly elucidated. Second, there are so many effects that it is illusory to be able to rebuild anything useful from their detailed analysis.

3.3 Accessing sensations is a real problem

Wittgenstein's point that "the assumption would... be possible —though unverifiable— that one section of mankind has one sensation of red, and another section, another", leaves us with little hope. This is among the reasons that lead him to consider that we cannot talk at all sensibly about privately sensed qualities. Some have argued even further that sensory quality has no objective reality.

These considerations sound like the final blow to our optimism for accessing, expressing, describing and analyzing sensations. Indeed, complete intrusion into a person's intimacy is impossible, even with all the person's help. How, then, can we proceed?

Apparently similar situations occur in other fields of science, and many formalisms have been proposed, such as fuzzy logic in the last years. New algorithms are being developed to tackle imprecise or uncertain data. But in this paper, the key is the ability to check for the adequation of a formal representation to the semantic aspect of our problem. This is outside the formalism itself. Consider the temperature scales for instance, Kelvin is a ratio scale, centigrade is an interval scale; calculating ratios is not allowed for the latter. Regarding taste, we do not know on which kind of scale we should work, and therefore which operations violate the "rules of perceptions". We can only agree to qualify as "red", or "sweet" a specific object.

4 FORMULATING A BEVERAGE WITH THE HELP OF AN AUTOMATED APPARATUS

Figure 2. The automated apparatus for the formulation of beverages.

To illustrate some ways of coping with this kind of difficulty, let us introduce a project of computer-aided formulation for a preparation limited to mixing liquid components. Let us say we intend to demonstrate the feasibility of an interactive, step by step, "optimal" formulation for naive users.

Possible compositions were defined by choosing the nature of seven components (low-alcohol beer, sparkling water, acid, sugar, two complex flavoring agents, and nutritive substances including vitamins). The concentration of beer and nutritive components were fixed. Sparkling water was added to complete the mixtures to a constant volume. The four last ingredients were allowed to vary within reasonable ranges.

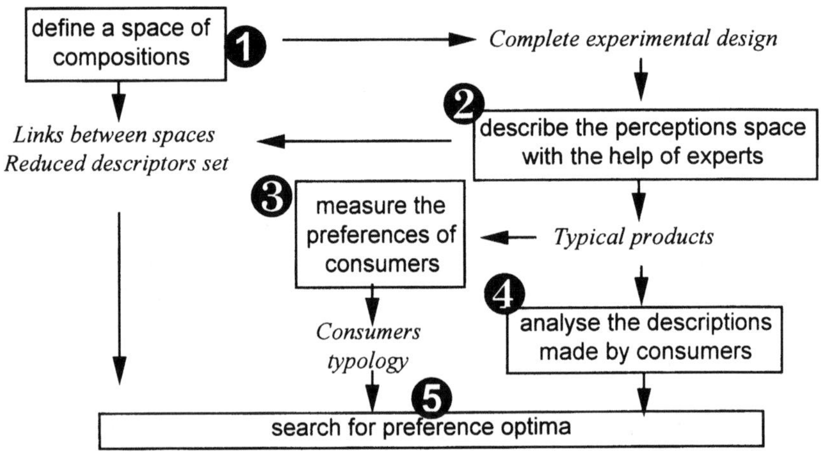

Figure 3. Formulating sensorily "optimal" beverages with the automated apparatus: the different steps.

This defines a composition space. The preference space is established by tasting and comparative rating. Perceptions must also be accounted for by a space if we want to be able to relate these three sets of information. The whole set of operation steps involved is shown in Figure 3.

Technically, a special mixing instrumentation was developed. The apparatus shown in Figure 2 comprises a series of reserve containers, each with an all-or-none valve, containers under high pressure with pneumatic jacks for distributing sparkling liquids, a cold unit. A computer, interfaced with a scale, the valves and jacks, controls the correct proportioning.

Establishing a language on gustatory sensations

An extensive factorial design (81 mixtures, 3^4) marks out the entire composition space.

A selected set of subjects is requested to find descriptors for the sensory properties. Working in pairs, they must succeed in identifying identical samples in glasses with a code number, through exchange of descriptions.

A first list of 113 descriptors is obtained. The aim here is *exhaustivity*; no perceivable difference should be neglected.

This list is not in fact utilizable as such and has to be reduced (using, for instance, the French AFNOR procedure), bearing in mind that we want to use it to write mathematical relations. We want terms associated with intensities (however rough) that are:

- *defined*, (understandable and memorable),
- *relevant and discriminating*, they allow for differences between mixtures. This is not the case with hedonistic terms that must, therefore, be banished. As they talk, the partners quickly realize that.
- *independent*, in particular, none should be calculable as a linear combination of the others; they should display null or weak correlation (Fig. 4),
- *unidimensional*, they refer to a "simple perception",
- *quantifiable*, they can be associated with a known numerical scale.

Descriptions must then be *repeatable*. How successful this stage is defines to what extent the status of a sensory metrology is reached. Measuring is the assignment of numerals to objects or events according to rules (Stevens, 1946). The final result is a grammatically poor tongue, but provided with primitive language properties.

It is established as belonging to a group of persons. Its use is restricted to a limited set of objects (here, the possible mixtures) and situations. It allows for communication of representations. What is exchanged seems limited, compared to what is felt by each. It is a living language. The odd measuring device, —a group of persons — vanishes if not sufficiently used, because the words of a dead language lose their meaning, their relation to reality and to people. Conversely, as they train, they become more expert in characterizing the products on which they are working. The only way to learn the language is to become part of the group and practice within its training methodology.

«odeur de levure» (yeast smell)	1.00				
«arôme mirabelle» (plum aroma)	-0.08	1.00			
«arôme pamplemousse» (grapefruit aroma)	-0.65	-0.43	1.00		
«acide» (sour)	0.13	-0.25	0.06	1.00	
«sucré» (sweet)	-0.06	0.33	-0.15	-0.47	1.00

Figure 4. Correlation matrix of the sensory descriptors [10].

<u>Putting those three spaces into relation</u>

A user, having tasted and rated the first sample, is asked to request modifications of the intensity of the sensations designated by the reduced descriptor set (Fig. 2, 4). It has been shown that naive (restrictively, French) users actually *understand* what is evoked by these names and manage to obtain recipes evolving in the direction they want. This means that the relevance of the language extends beyond its initial frontiers. This openness is expected from any language.

The requests to strengthen or soften the different sensations are to be translated back into a composition. This is far from trivial: sweeter, for instance, can generally be obtained either by adding sugar or by removing acid. Many aromas reinforce or diminish the sweet sensation. The translations from the perception space to the composition one are calculated using both linear regression and neural networks (for nonlinear modeling) trained by back-propagation. The training set was a fraction of the products of the factorial design, associated with their descriptive profiles (the remainder was kept to test the learning procedure). Each time the computer calculates a new mixture that fits the requirements. Exploration thus evolves interactively by a series of trials.

5 How does Exploration Fit into the World of Taste Optimizing?

Our initial purpose was to make some room for science in cookery. So far we have noted that some scientific methods can be applied to describe sensations. It is tempting to go beyond our interactive process and to run an optimization algorithm in the perception space.

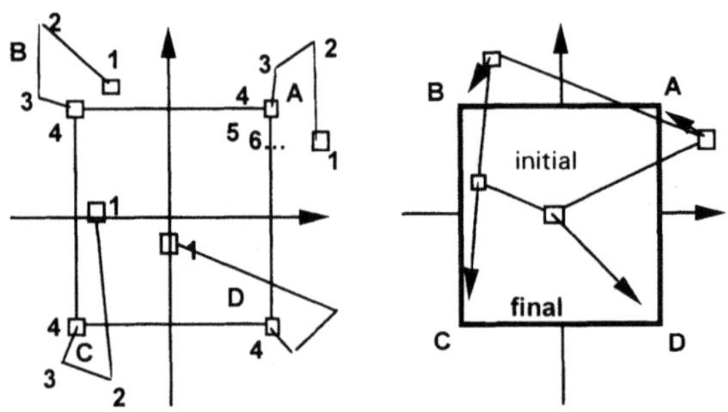

Figure 5. Evolution of descriptions over sessions (See text for comments).

Figure 5 displays the generic aspect of a first factorial plot, as can be obtained by principal component analysis over the descriptive profiles of four products, *A, B, C, D*. A judge is requested to describe each; he repeats this operation for several sessions, denoted *1, 2, 3...* The points representing the descriptions of a product, e.g. *A*, descend along the plane during the first sessions. Eventually, a steady state is reached and the judge is able to give stable answers; he can be a good measuring device. There is a learning period which allows the taster to progressively situate each product among the others. The relations between *A, B, C* and *D* can be strongly modified meanwhile. This behavior is seen each time the taster is not already an expert, e.g. an oenologist for wine. For any kind of foodstuff, a huge majority of people is, therefore, in the initial amateur state; they are the actual market.

Designing an optimal recipe for them cannot be done by relying on their descriptions, because they are so unstable that their preferences cannot be located (*A* is preferred; is it on *1, 2*... or elsewhere?) No optimization method is applicable because all require spatial information that does not exist in the present case; it is created when people become experts, but at the same time, they cease to be statistically representative samples of the market.

The usual workaround is to use an expert's description and to ask for simple preferences (like *A>B?*) between a few products (two to four, more being too confusing) in a market survey. Meanwhile, we jettison a huge amount of information people would like to express. However contingent it might appear, the wandering "interactive optimization" seems the only possible one as long as you want naive subjects to act directly upon the modifications of a recipe.

What we are faced with is a space irreversibly changing under measurement. As Tomas knew in Kundera's novel *The Unbearable Lightness of Being*, "Did he want her to come or did he not?", there is no way to try both experiments. History cannot be optimized. And our relation to food has a strong historical component.

CONCLUSION

Contrary to physics, where the observer in the measuring process becomes important for the theory only when wave function reduction falls on Schrödinger's cat paradox, sensory analysis has to make do from the beginning with the psychology included in its measuring device.

Representation can be seen as the core of science, even of knowledge in general. Modeling something means transforming a space to achieve a representation that will be easier to use, where elements can be easily combined. We always look for spaces for which both elements and operators will seem isomorphic to the phenomenon we witness. In cookery, we do have some elements —sensations—, but determining the rules as to how to combine them mentally so that they will express the result of a recipe is extremely difficult. Flavors resist abstraction and mental composition. That may be what makes art the correct way to address cookery.

KEYWORDS

Representation, sensory analysis, taste, optimization.

ACKNOWLEDGMENTS

This work was partly supported by the Direction Générale de l'Alimentation, under grant 90 LR 20. I would like to thank Professor J.-E. Dubois and Bernice Dubois for their interest and helpful remarks.

REFERENCES

[1] Afnor, Analyse sensorielle, Méthodologie, Méthode d'éveil à la sensibilité gustative, NF V 09-002 (1989),
Initiation et entraînement à la détection et à la reconnaissance des odeurs. NF V 09-006 (1989)
Recherche de descripteurs pour l'élaboration d'un profil sensoriel. NF V 09-21 (1990)

[2] Amerine M.A., Pangborn R.M., Roessler E.B., Principles of sensory evaluation of food, Academic Press (1965)

[3] Bardot I., Méthodes d'optimisation appliquées à la formulation: cas de la formulation sensorielle, PhD. Thesis, ENSIA, Massy (1993)

[4] Bardot I., Martin N., Trystram G., Bochereou L., Rogeaux M., Hossenlopp J., A New Approach for Formulation of Beverages, Part II : Interactive Automatic Method, Lebensmittel-Wissenschaft und Technologie (1994)

[5] Blanc-Mouchet J. (Ed.) Odeurs. Autrement, Paris (1992)

[6] Heyd B., Bardot I., Trystram G., Hossenlopp J., Automated Formulation of a Beverage Based on the Optimization of Sensory Properties, Food Quality and Preferences, Vol 6, 1, pp1-6 (1995)

[7] Humphrey N., A History of the Mind, Vintage (1993)

[8] Mac Leod P., La perception des aliments nouveaux, IAA, 367:69, May (1991)

[9] Mac Leod P., Circuits nerveux de la préférence alimentaire et du plaisir, in: Plaisir et préférences alimentaires, Giachetti I. (Ed.), Polytechnica (1992)

[10] Martin, N., Exploration d'un espace de perceptions et d'un espace de préférences, recherche d'optima en analyse sensorielle, PhD. Thesis, ENSIA, Massy, (1993)

[11] Martin N., Bardot I., Hossenlopp J., Trystram G., Rogeaux M., Bochereou L., A New Approach for the Formulation of Beverages, Part I: Sensory Exploration of a Composition Space by Trained and Consumer Panelists, Lebensmittel-Wissenschaft und Technologie (1994)

[12] Moskowitz H. R., (Ed.), Product testing and sensory evaluation of foods, Food & Nutrition Press, Inc., Westport, Connecticut (1985).

[13] SSHA, ISHA, Évaluation sensorielle, Manuel méthodologique, Lavoisier, (1990).

[14] Stevens S. S., Measurement, Statistics, and the Schemapiric View, Science, Vol. 161, 3844, pp849-856 (1968)

[15] Stone H., Sidel J. L., Sensory Evaluation Practices, Academic Press, Inc.(1985)

[16] Süskind P., Das Parfum, Diogenes Verlag ag, Zürich, 1985. Perfume, Transl. John E. Woods, Hamish Hamilton (1986)

THE INTEGRATION OF SCIENTIFIC DATA INTO AN ARTIFICIAL INTELLIGENCE SYSTEM

Adam C. WINSTANLEY and F.J. SMITH

Artificial Intelligence and Applications Research Group Department of Computer Science, Queen's University, Belfast BT7 1NN, Northern Ireland, UK

ABSTRACT

Part of the knowledge of a physical scientist or engineer is an understanding of the laws of science represented by formulae and constraints found in the conventional text books of science and engineering. This kind of knowledge can be represented naturally in an object-oriented paradigm and can become part of an artificial intelligence system for solving quantitative problems in the physical sciences. However, to solve real problems such an artificial intelligence system can only work if it includes a database of fundamental constants and of the scientific data pertinent to the domain covered by the system. It must also be able to manipulate the system(s) of units used in that domain. When all these are integrated together the resulting system provides a powerful tool to aid a scientist or engineer. This paper describes the data modeling in such an artificial intelligence system.

RESUME

Une partie de la connaissance d'un chercheur physicien ou d'un ingénieur est la compréhension des lois de la science représentées par des formules et des contraintes des manuels conventionnels scientifiques et techniques. Ce type de connaissance peut être représenté tout naturellement par un "paradigme à base d'objets" et peut devenir un sous-ensemble d'un système d'intelligence artificielle (IA) pour résoudre quantitativement des problèmes des sciences physiques. Cependant, des problèmes réels ne peuvent être résolus par un tel système IA que si le système comprend une base de données de constantes fondamentales et de faits scientifiques pertinents aux domaines couverts. Lorsque tous ces aspects sont bien intégrés ensemble, le système obtenu se comporte comme un outil puissant d'assistance au chercheur et à l'ingénieur. Cet article décrit la modélisation des données dans un tel système IA.

1 INTRODUCTION

One of the main areas of research in the Artificial Intelligence Research Group at Queen's University has been the storage and manipulation of the complex forms of data used in science and engineering. One result has been the construction of an artificial intelligence system known as QPS (Quantitative Problem Solver) [1]. QPS can be described as a meta-tool for building quantitative problem solvers for particular domains in science and engineering. It consists of three main parts :

- A problem solving engine
- A graphical user interface
- A knowledge base of scientific data which can be customized for different domains.

QPS runs using the XWindows or SunView environment on SUN Sparc Work-stations. It was constructed using object-oriented design techniques [2]. A previous paper [3] described the architecture and problem solving strategy used by QPS. Here we described its data modeling capacities.

2 The Nature of Quantitative Knowledge

The knowledge required to solve a quantitative problem can be divided into three main categories: that which is applicable to all problems (referred to here as *universal* knowledge), that which is relevant to a particular domain of engineering (*domain* knowledge) and transient data used in solving a particular problem (*problem-specific*).

2.1 Universal

For any quantity to be meaningful, it is necessary to know the units in which it is measured. Despite the widespread adoption of S.I. units, the imperial system (and other systems) of measurement are still in use in many areas of engineering. Knowledge of the relationship between the various units of measurement can therefore be considered necessary for the solving of all quantitative problems.

Another type of universal knowledge is the nature and values of the fundamental constants such as the speed of light (**c**), the charge on an electron (**e**) and the value of the mathematical constant π.

2.2 Domain

The solutions to problems in different fields of engineering tend to depend on different sets of knowledge. For example, electrical engineers use formulae relating current, voltage and resistance whereas structural engineers deal with tensile stress and strain and so on. This domain-specific knowledge can be subdivided as follows :

- the characteristics of geometrical shapes, for example cylinders, cubes, bars
- the physical properties of materials (the resistivity of copper, the specific heat capacity of aluminum alloy)

- the formulae that relate quantities, usually expressed as mathematical equations (for example tensile stress is given by the formula $\sigma = F/A$ where F is the applied force and A is the cross-sectional area). Additionally, some formulae have constraints on their appropriate use.

2.3 Problem-specific

A quantitative problem is specified in terms of the values of certain physical or state variables, for example :

What is the tensile stress on an aluminum cylinder 0.5m long that will produce a change of length of 0.1mm?

The physical variables involved in this problem are :

original length	$l_0 =$	0.5m
change of length	$\delta l =$	0.001m
tensile stress	$\sigma =$	?Pa

3 THE ORGANIZATION OF KNOWLEDGE WITHIN QPS

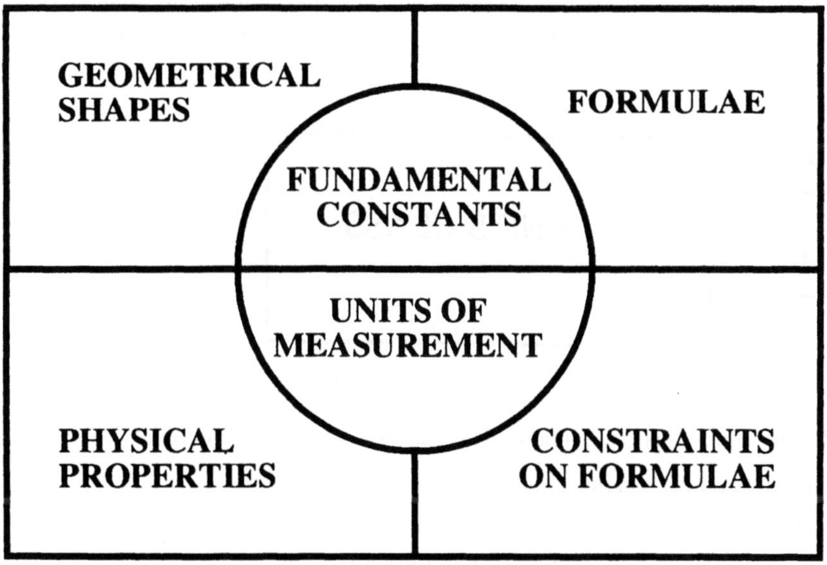

Figure 1. The organization of knowledge within QPS

QPS reflects these three divisions of quantitative knowledge in the organization of its knowledge base (Fig. 1). Fundamental constants and units of measurement are built into the system providing a core of basic universal knowledge for problem solving. To create a problem solving tool, data specific to the target domain are added. In particular, this includes databases of the values of the physical properties of materials and geometrical shapes. Also added are the formulae relating different quantities, together with any constraints that are placed on their use. To solve a particular problem, the user provides the names or symbols of the known state variables involved and their values.

Problem solvers built using QPS can be considered expert systems, with formulae (and their constraints) acting as inference rules for the scientific data of constants, units of measurement and physical properties. The data on fundamental constants are stored in a simple table. Those of the other two are described in the following sections.

4 Units of Measurement

In QPS, all data are automatically converted to base SI units before problem solving begins. After deriving a solution, the value of the goal quantity is converted to the units specified for the answer. These conversions are accomplished using a software object built into QPS which contains a database of units and conversion formulae (Fig. 2). The database is extendible by users if they wish to use more unusual units.

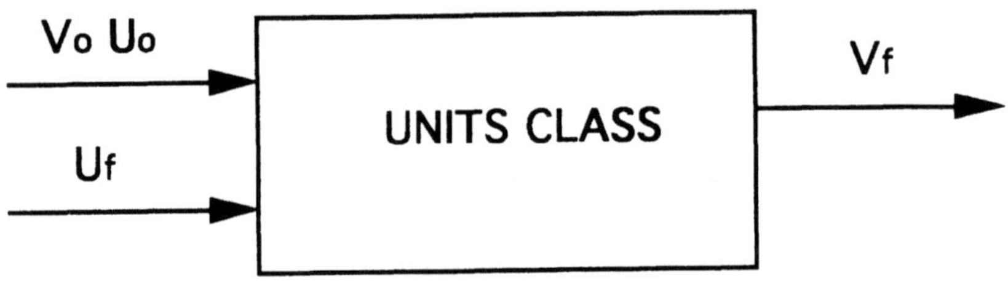

Figure 2. The Units Class showing original and final values (**V**) and units (**U**)

The in-built units object handles :

- base SI units.
- derived SI units: Expressible in terms of base units
- standard S.I. Prefixes, e.g. centi (c), kilo (k)
- imperial units
- conversion formulae between units, e.g. T°C= (T+273.15) K

5 PHYSICAL PROPERTIES

The data available from reference books on the physical properties of materials are vast and complex. For example, the elastic modulus of an item can depend on the precise material it is made from (e.g. 6061 Aluminum Alloy), the form of the specimen (bar, sheet, plate) and its environment (e.g. temperature). QPS has an internal object-oriented database that can store and manage property data but using this has a number of disadvantages :

- Large data sets can be time-consuming to input.
- Object-oriented modeling of complex data can be difficult for non specialists.
- The integrity of the data is hard to assure.

QPS, therefore, has been extended with the ability to interface with other systems such as the hierarchical database management system *M/Vision* [4]. As well as providing storage and retrieval facilities for complex user-defined data models, M/Vision also has available large verified data banks of material properties of metals, ceramics and plastics.

The use of external databases during problem solving is largely transparent to the user. Once set up using specification files, QPS automatically knows where to obtain each particular class of data. Also, different levels of detail are supported. For example, if the material is specified as *Aluminum* (rather than a more specific description such as *2106 Aluminum Alloy*), QPS itself will select a representative sample.

6 AN EXAMPLE PROBLEM

In order to illustrate the use of QPS, the solution to the problem mentioned in section 2 was found :

> *A cylinder of Aluminum Alloy is 0.5m long. What amount of tensile stress on it will produce a change of length of 0.1mm?*

The problem can be specified as a list of input values, and a goal value :

material =	2014 Aluminum Alloy
geometry =	cylinder
original length =	0.5m
change of length =	0.1mm
goal tensile stress =	? Pa

Figure 3 shows the problem as input into the SunView user interface for QPS. The interface is designed so that problems are specified compactly using the symbols, superscripts and subscripts familiar to engineers and scientists. Once the *solve* button is selected, QPS verifies the problem specification for consistency of units and dimensions and then calculates the result. It does this by selecting appropriate formulae and retrieving data on constants and properties from its knowledge base [5].

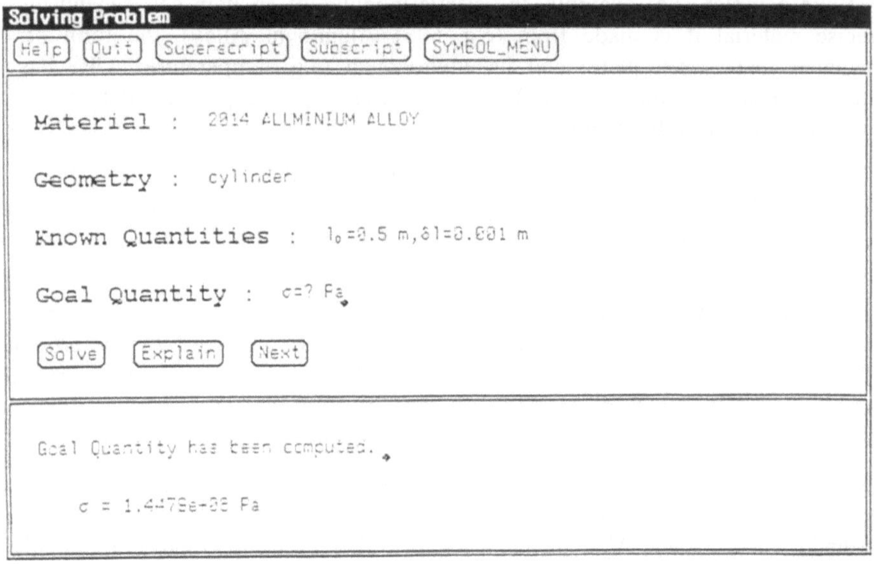

Figure 3. Specification of a problem in QPS (Sun View version)

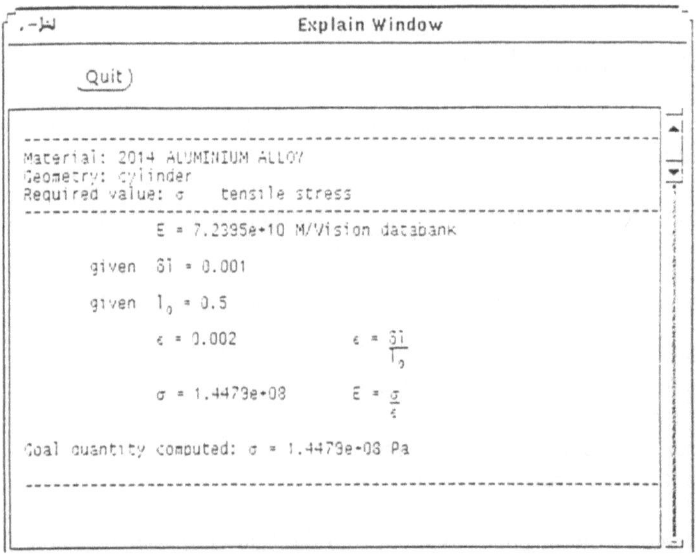

Figure 4. Exploration of solution path by QPS (XWindows version)

In order to explain the solution path used, the system can display the values obtained at each stage in the calculation (Fig. 4) together with the source of that value (i.e. whether it was given by the user, obtained from a database or calculated with a particular formula).

CONCLUSION

In trials with industrial partners, QPS has proved to be a useful tool, solving many common types of quantitative problems correctly and in a much shorter time that if done by conventional means. However, these trials have also shown up deficiencies in the system which are now being addressed.

Formulae:

The direct manipulation of simple formulae, including those involving differentials and integrals, is implemented within QPS. The modeling of more complex formulae, such as those involving partial differentials (as in fluid mechanics) or multiple integrals is being accomplished using external software functions either custom written or retrieved from software libraries. The object-oriented design of QPS makes it amenable to extensions of this kind.

Fuzzy data:

Uncertainty in input data or that representing properties of materials is being tackled in three ways. Facilities are being developed for specifying :

- The *precision* of real numbers supplied by the user and using these to calculate the precision of the goal value.
- *intervals* and ranges of data values
- *fuzzy numbers*, modeled using low, high and median values

Due to the object-oriented design of QPS, these extensions will all be able to use the existing problem-solving algorithms.

User interface:

Problems are currently specified textually. We are improving the user-interface using X Windows so that problems can alternatively be specified graphically. This involves the display of 2D and 3D shapes labeled with appropriate quantity symbols. The user will specify the problem by interacting with the diagram to provide the input values.

Material selection:

We plan to extend QPS to allow it to intelligently select candidate materials for components [6]. By using the knowledge base of data and formulae, the materials that match specified criteria can be selected and suggestions made for the optimum material and optimum design for the component.

ACKNOWLEDGMENTS

The following people contributed to the work described here: M.V. Krishnamurthy, Shakti Tripathy, Paul Sage, Jaron Collis, Stéphanie Chalmé, Joanna Brennan.

KEYWORDS

Quantitative problem solving, quantitative knowledge, units of measurement, material properties databases, M/vision.

REFERENCES

[1] Krishnamurthy M.V., A Tool for Quantitative Problem Solving in Science and Engineering, PhD Thesis, Queen's University, Belfast (1993)

[2] Meyer B.: Object-Oriented Software Construction, Prentice Hall (1988).

[3] Krishnamurthy M.V., Smith F.J., The Representation of Textbook Knowledge on Physical Science, New Data Challenges in Our Information Age, Proceedings 13th International CODATA Conference, A-39-46 Oct.(1992)

[4] PDA Engineering: M/Vision Materials System Builder Users Guide and Reference (1993).

[5] Krishnamurthy M.V., Smith F.J., Problem Solving in Engineering Using Quantitative Knowledge, Knowledge based Systems Journal (1994).

[6] Winstanley A.C., Smith F.J., The Intelligent Selection of Materials from a Design Specification, this volume (1994).

MULTIMEDIA INFORMATION SYSTEM FOR MACHINING OF ADVANCED AEROSPACE COMPOSITES

Horris C. LEUNG, Ian R. PASHBY and W. Y. TSUI

Department of Engineering, University of Warwick, Coventry CV4 7AL, U.K.

ABSTRACT

A unique experience in the material processing community of developing an information system to manage the data generated by a research program on the machining of advanced aerospace composite materials is discussed.. The knowledge generated by the research, due to its very nature, is normally expressed in terms of various plots on the numeric data captured from numerous experiments, and is usually associated with textual and imaging information. In our program certain phenomena can only be well explained by motion sequences or video clips. The paper discusses our approach for the purpose of interesting users ranging from material scientists in research laboratories to machinists on the shop floor.

RESUME

On discute une expérience unique de la communauté concernée par l'élaboration des matériaux en vue de développer un système d'information pour gérer les données par un programme de recherche sur la préparation de matériaux composites aérospatiaux. La connaissance engendrée par la recherche en fonction de sa spécificité est normalement exprimée en terme de blocs de données numériques captées à partir de nombreuses expériences et est usuellement associée avec des informations textuelles et des informations associées à des images. Dans le programme, certains phénomènes ne sont valablement expliqués que par des séquences de déplacements ou vidéoclips. Dans ce travail, on discute notre approche conduite dans l'intérêt des utilisateurs allant du chercheur en matériaux jusqu'aux machinistes de l'atelier.

1 INTRODUCTION

Machining of modern materials is increasingly difficult following the advances of material processing and manufacturing, conventional machining processes have been found inappropriate or inapplicable in many cases as a result. The Machining of

Composites Program, funded by the Department of Trade and Industry (DTI) and in collaboration with a consortium of seven companies in Britain was set up therefore to do research on the machining of a number of advance composite materials. The materials studied by the program fall into two broad families, the Polymer Matrix Composites (PMC) and Ceramic Matrix Composites (CMC). The manufacturing methods of these materials are such that the distribution of fibers within the material is non homogeneous in all orientations. This will obviously have significant effects on the processibility of the materials. Consequently pulsed Nd-YAG solid state laser and CO_2 laser are proposed to machine the materials under investigation. By studying the effects of different settings of parameters such as laser power, type of assist gas and others, we aim to optimize the material removal rate and the processed surface quality [1,2].

It was recognized in the planning stage of the program that the amount of data to be generated would be enormous; a new approach in managing and disseminating the information is necessary if interested parties are to obtain maximum benefit from the program.

The paper is organized as follows. We describe in Section 2 the emerging models of multimedia information systems and the existing applications developed for different purposes. Section 3 is a discussion on the various types of data entities made available by the program and how they are normally used by material scientists. Following this in Section 4 we define the users and functional requirements and analyze the application scenarios, and derive as a result the functional requirements of the information system. Section 5 focuses on our approach towards developing the system. How the idea is being implemented will be presented in Section 6. The system was demonstrated to different groups of material processing and machining experts. We conclude the paper with a discussion of their observations and the implications for future work.

2 Multimedia Information Systems - Review on the State of the Art

In recent years multimedia and multimedia information systems have drawn a lot of attention from database researchers and database practitioners.

2.1 Multimedia - a definition

Extending this idea of multimedia for information management, Grosky [3] proposes that, in addition to the commonly mentioned entities like images, audios and videos, any multidimensional array of numbers derived from various sensors should be classified as multimedia data entities. Therefore, 3D data collected by a Coordinate Measuring Machine (CMM) in a manufacturing environment, spectrums generated by an ultra-sonic scanner in a laboratory, or ultimately scientific data generated from experiments through the mediation of various sensors may well be considered as multimedia data entities.

2.2 Multimedia information system - the essentials and issues

The need for a multimedia information system to support scientific applications is a direct consequence of the limitations of the traditional Database Management Systems (DBMS). Such limitations can generally be summarized as limited support of diverse data types and the corresponding processing operations [4]. It is quite obvious that given the possible variety of data entities constituting multimedia, the scope of research expertise involved is enormous. Generalizing the case with the given definition of multimedia data entities, this process can be formalized as follows [3,5,6].

1. Decoding the multimedia data entities residing in the database;
2. Extracting features that best represent or describe each of the data entities (e.g. color and body style in the example given);
3. Indexing and referencing the features;
4. Selecting the appropriate multimedia data entity/entities that match the current query.

The current solution to implementing multimedia information systems is to opt for manual or semi-automatic alternatives [3]. Furthermore, it is also understood that an appropriate database model and a query language are yet to be researched to support the requirements of a multimedia information system [5, 6].

2.3 Conceptual models

It is generally agreed that multimedia information system design is still in its infancy [3].

Recent developments in relational databases allows for the storage of Binary Large Object (BLOB)[7] and consequently, storage of multimedia data entities within relational databases becomes possible.

Currently, almost all research effort points towards the use of hypertext to tackle the problem [8, 9,10 ,11].

Hypertext is defined as text in electronic form that takes advantage of the interactive capabilities [12].

While the concept of multimedia information system is still evolving, a complete systematic review on every research perspective is clearly out of the scope of this paper. However, before finishing this section, the authors would like to emphasize two further aspects which will have significant impact on the future of multimedia information system. Firstly, the use of artificial intelligence is vital. Secondly, inspired by the fact that databases of various kind are naturally distributed, in both a logical and physical sense, Bertino et al [14] argue that knowledge bases and multimedia databases should be integrated to form a kind of intelligent information management system. As the work of Bertino et al is mainly on integrating logically distributed databases, Merlet et al [15] propose to integrate physically distributed databases by the object oriented paradigm.

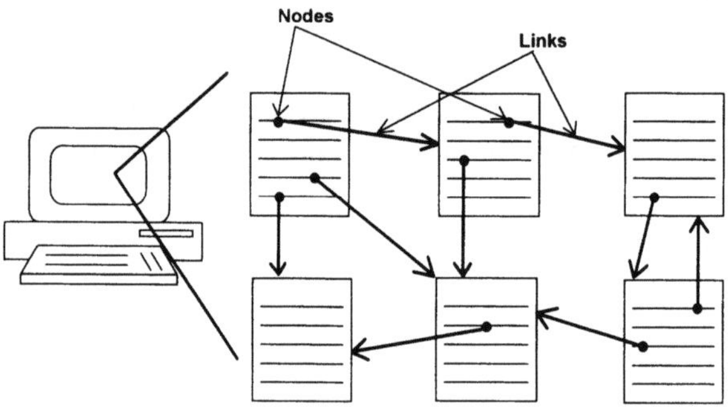

Fig. 1 Nodes and Links of Hypertext

2.4 Applications

In general engineering, Kaye et al [16] in NASA initiated a STI (Scientific and Technical Information) program which aimed to provide scientists and engineers with an information environment through which a large body of information becomes easily accessible. This information, generated by numerous NASA research programs and usually multimedia by nature will be presented to the end users by the concept of hypermedia.

In a less ambitious program, McMahon et al [17] described the work on providing engineers with what they called informal design data. Sources of these data are quite diverse from product literature to reference textbook.

3 DATA ENTITIES

The various type of data entities generated by the program and how they relate with one another is a major factor determining the design of the multimedia information system. These are discussed as follows:

- *Alphanumeric data*
  This type of data are the records of quantifiable measures on various performance indices, for example the effect of pulse intensity variation in using Nd-YAG laser for the cutting of Glass Matrix Composites as shown in Figure 2

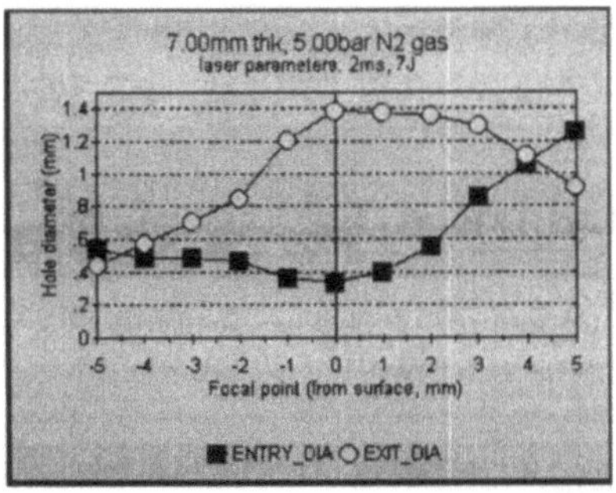

Fig. 2 Sample plots of quantifiable data

- *Photographic images*
  These are usually the images recording the results of physical analysis on the reaction of the materials in response to the changes in cutting conditions. Figure 3 shows an example of an image from the Scanning Electron Microscopy (SEM).

Fig. 3 SEM analysis identifying a thermally induced micro-cracking in the material

- *Video sequences*
  This type of data entity is used to explain phenomenon that changes in relation with time.

- *Technical reports*
  These are the reports distributed to the industrial collaborators.

- *Publications*
  These are papers published in journals or conferences based on the knowledge generated by the program.

- *Literature*
  This refers to the publications or technical reports compiled by other research groups working in the related areas of the program. They are

stored normally as scanned image from the original document that we manage to access.

4 USERS AND FUNCTIONAL REQUIREMENTS

By a hypothesis on the application scenarios on both environments, we are able to draw up a list of functional requirements to guide one through the system design and development.

4.1 Application scenarios

Users of such systems in a research laboratory are primarily material scientists researching possibly on the use of one of the machining processes available from the information system but on a different material than that investigated. The situation is quite different in an industrial environment. Usually, a machinist is given a material along with a machining process that is largely predefined based on the facilities available. Consequently, he will search in the information system base on some performance criteria, for example, surface finish and cutting time. As an output from the system, he would expect information which contains advice and data in order to obtain some form of optimum cutting condition.

In summary, the distinction of these two modes of application can be characterized as follows :

- The information required by machinists in the industry is quite specific, based on which he can solve a particular problem;
- Scientists in a research environment need information on most aspects of the problem area to assist them in the problem solving process;

5 THE INFORMATION INFRASTRUCTURE

It is relatively straightforward that each type of data entity listed in Section 3 needs a data repository. The problem as discussed before is to decide which type of database model to adopt. Since a relational database is still the dominant database model in almost every sector of information system application, it is adopted therefore as the database model for our data repositories. Figure 4 shows the repositories constituting the information system and their inter-relationships.

It is quite important that these repositories are designed so that they are related to one another in order to support the query processing in a relational database model. From the diagram, it can be observed that the alphanumeric data repository plays the most important role. On one hand, it stores all the quantifiable data Multimedia Information System generated by the experiments, and on the other hand, it provides the links to the rest of the repositories, either directly or indirectly.

The material types and parameters like cutting speed, assist gas pressure, are normally used to establish the relations between the alphanumeric data repository and that of photographic images. The relations from alphanumeric data to videosequences are established in a similar fashion.

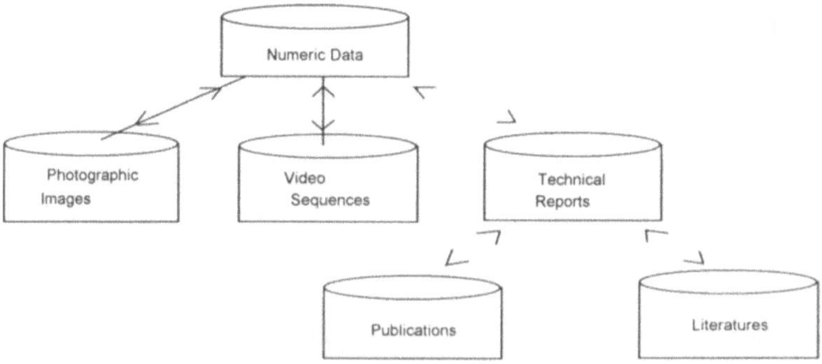

Fig. 4 Multimedia repositories and the inter-relationship

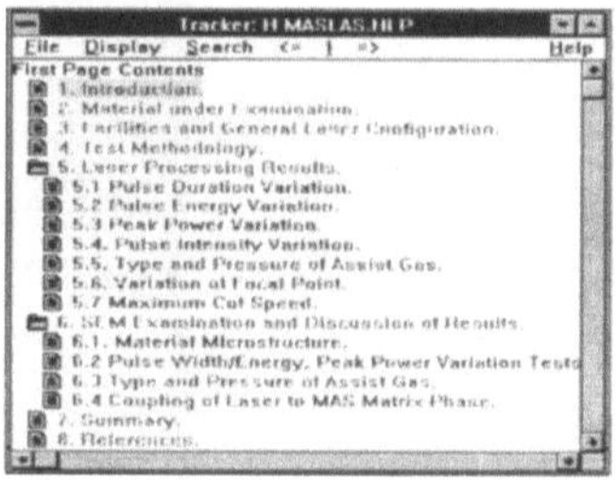

Fig. 5 Content page of a technical report converted into hypertext

6 THE IMPLEMENTATION

A personal computer operating under MS Windows has been chosen as the platform to implement the information system.

Currently, the information infrastructure depicted in the previous section along with a few information navigation mechanisms have been built. These mechanisms are summarized in Figure 6.

The first three blocks in Figure 6 allow the users to specify the parameters for information retrieval. The users will then be given the choice of visualizing the information based on the type of analysis that our research group has performed throughout the program. For instance, the analysis on the Nd-YAG machining normally includes:

- Pulse duration variation

- Pulse energy variation
- Peak power variation
- Pulse intensity variation
- Assist gas type
- Assist gas pressure
- Variation of focal point
- Maximum cut speed

Figure 7a to Figure 7c demonstrate some of the results that a user can retrieve during a specific search.

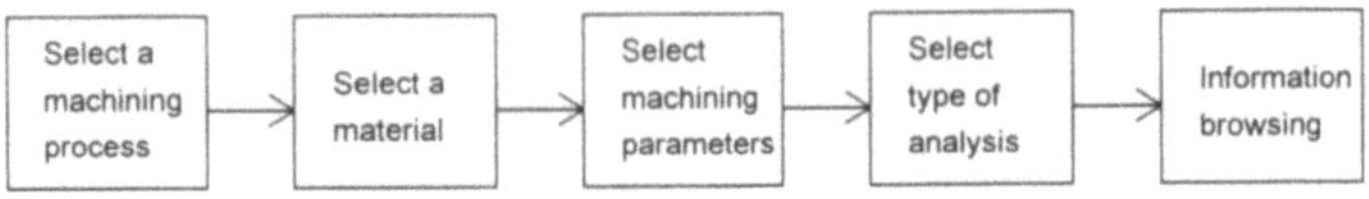

Fig. 6 : Currently supported information retrieval mechanisms

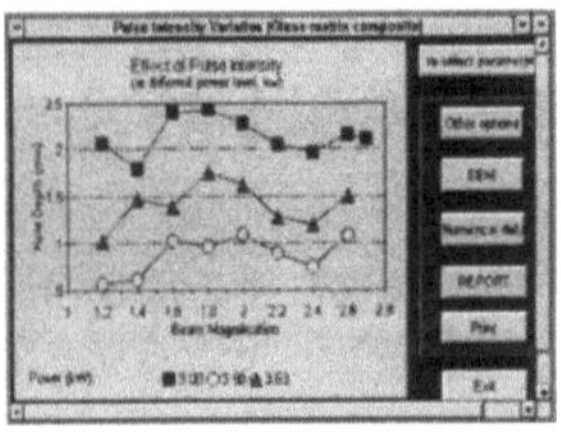

Fig. 7a : Numeric data visualized based intensity variation analysis.

Fig. 7b : An image showing the pulse evidence of microcrystallization. Note also in the diagram the comments and the facilities available to search and browse through the other images resulting from the search.

Fig. 7c : The section of technical report explaining the pulse intensity analysis and corresponding results is displayed with a hypertext browser. The information residing in the literature and publication repository can be accessed through a hypertext link within this window.

7 Results and Discussions

The system in its current form has been demonstrated to several groups of material processing and machining experts so as to refine and adjust the development direction. The feedback has been positive and fairly encouraging so far. This is partly due to the fact that it is the first time such an integrated information system is developed in the community. More importantly, however, it is a user friendly tool that delivers quality information efficiently and effectively.

It is nevertheless being suggested that the current system be oriented more towards research rather than industrial application, as a majority of the functions supported now is for information navigation and discovery

Conclusion

A multimedia information system has been developed for managing the results of a laser machining research program. In particular, the user friendliness, the speed, quality and availability of information are most valuable. Further work needs to be done in order that a equivalent level of support be available for industrial use.

Acknowledgment

The authors would like to thank the DTI and the consortium of seven companies who agreed that this work be published. The work forms part of a research project under DTI's MTTPP initiative which is carried out in the University of Warwick.

Keywords

Information technology, hypertext, multimedia, electronic publishing, databases, world-wide web.

References

[1] Tuersly I. P., Hoult A. P. , Summary of Systematic Machining Trials of SiC/Pyrex GMC Material Using YAG Laser, MTTP Program Report, March (1993), (Unpublished technical report, University of Warwick)

[2] Barnes S., Fenoughty K., The Cutting of Carbon Fibre Epoxy Composite with a CO2 Laser, MTTP Program Report, June (1993) (Unpublished technical report, University of Warwick).

[3] Grosky W. I., Multimedia Information Systems, IEEE Multimedia, pp 12-24, Spring. (1994)

[4] Chakravarthy S., Krishnaprasad V., Tamizuddin Z., Lambay F., A Federated Multimedia DBMS for Medical Research: Architecture and Functionality, Technical Report UF-CIS-TR-93-006 University of Florida, Computer and Information Science, pp 1-12 (1993)

[5] Friesen O., Golshani F., Multimedia Information Systems: The Confluence of Multiple Technologies, ICCS/ISITA'92 Conference Proceedings, Communications on the Move Singapore, Vol 2 pp 969-973 (1992)

[6] Mallett J., Manning C., Multimedia and Database Design: A Discussion of Database Technology and Its Use in Multimedia, Journal of Document & Text Management, Vol 1 N° 1 pp 75-84 (1993)

[7] Shetler T., Birth of the BLOB, BYTE, pp 221-226, Feb (1992)

[8] Cavallaro U., Paolini P., Christodoulakis S., Dallas C., Enotiadis A., Prioa S., Schiff J., Schuler W., HIFI: Hypertext Interface for Information: Multimedia and Relational Databases, The Electronic Library Vol 11 N° 2 pp 65-71 (1993)

[9] Frei H.P., Schauble P., Designing a Hypermedia Information System, Proceedings of DEXA'91 International Conference on Database and Expert Systems Applications, pp 449-454 (1991)

[10] Leigh W., Ragusa J., Gilliam M., An Architecture for a Heterogeneous Information Environment, Journal of Computer Information Systems, pp 42-49 Fall (1993)

[11] Bieber M., Providing Information Systems with Full Hypermedia Functionality, Proceedings of the 26th Hawaii International Conference on System Sciences, pp 390-400 (1993)

[12] Conklin J., Hypertext - A Survey and Introduction, IEEE Computer, Vol 20 N° 9 pp 17-41, (1987)

[13] Yankelovich N., Haan B.J., Meyrowitz N.K., Drucker S.M., Intermedia: The Concept and the Construction of a Seamless Information Environment, IEEE Computer, Vol 21 N° 1 pp 81-96 (1988)

[14] Bertino E., Damian M., Paolo R., Spampinato L., An Advanced Information Management System, Proceedings RIDE-IMS'93 Research Issues in Data Engineering: Interoperability in Multidatabase Systems, pp 61-64 (1993)

[15] Merlet J.F., Nadalin C., Soutou C., Lapujade F.R., Toward a Design Method of a Multimedia Information System, Proceedings of 1993 IEEE International Conference on System, Man and Cybernetics, Vol 4 pp 367-372 (1993)

[16] Theodoropoulos G., Loumos V., Filippakis V., A Multimedia Relational Database Program for Parasite Identification, Computer Methods and Programs in Biomedicine, Vol 39 pp 297-301 (1993)

[17] Kaye K.M., Kuhn A.D., Hypertext and Hypermedia Systems in Information Retrieval, Bringing Down the Barriers to Information Transfer (AGARD-CP-505), pp 6/1-6/5 (1992)

[18] McMahon C.A., Pitt D.J., Yang Y., Williams J.H.S., An Information Management System for Informal Design Data, AISB Quarterly, pp 54-59 (1993)

REASONING ABOUT CORROSION FROM CASES INSTEAD OF RULES

Charles P. STURROCK[1,2] and Walter F. BOGAERTS[2]

[1] National Institute of Standards and Technology, Gaithersburg, Maryland, 20899 USA
[2] Katholieke Universiteit Leuven, Dept. of Materials Engineering, 3001 Leuven, Belgium

ABSTRACT

There are no rigorous, generalizable models governing corrosion behavior analogous to, for example, the Navier-Stokes equations of fluid mechanics. Much of what is known about corrosion has been derived from case histories of actual failures and laboratory work. In this respect, the practice of corrosion failure analysis is very much like that of medical diagnosis: although we can simulate real-world phenomena in the laboratory, we tend to rely heavily on actual case data. Thus corrosion has served as an excellent domain in which to develop knowledge-based systems(KBS). Two recent surveys point to 57 KBS projects. Virtually all of them infer conclusions from rules distilled from previous experience.

RESUME

Il n'y a pas de modèles généralisables et rigoureux qui gouvernent le comportement de la corrosion. Les faits connus de la corrosion sont identifiés souvent à partir de cas de faiblesse bien connus, autant qu'à partir de données recueillies dans les expériences de laboratoire. Bien que nous puissions simuler des phénomènes du monde réel en laboratoire, nous avons tendance à nous reposer sur l'enseignement des données des situations connues. Comme la médecine, la corrosion a pu être un excellent domaine de développement des systèmes à base de connaissances "SBC". Deux études récentes ont mis en évidence l'existence de 57 projets SBC en corrosion. Virtuellement tous ces systèmes tirent des conclusions par inférence à partir des règles déduites d'expériences .

1 INTRODUCTION

We have recently accumulated and analyzed 84 cases of documented corrosion failures in the process industries. Six different failure mechanisms are represented among the 84 cases. One objective of the research has been to determine whether we could predict the operative corrosion failure mechanism by identifying patterns in the case data. Unfortunately, there are no standards for reporting cases of corrosion failures, so the available case data is disparate, and does not lend itself to statistical analysis or pattern recognition using artificial neural networks. We represent the cases of corrosion failures as points mapped onto a hypercube with a diagonal of unit length. The dimension of the hypercube is determined exclusively from the case data. Distances within the experiments calculate them according to a dissimilarity metric derived from geometric and intuitive arguments. Our initial findings suggest that similar corrosion phenomena cluster within specific regions of the multidimensional feature space circumscribed by the case data: reasoning on purely phenomenological grounds, we are able to predict the operative corrosion mode with as much as 96.4% accuracy, as compared with a random guess rate of 16.7%. These results demonstrate a very promising potential for the use of analogical reasoning in the corrosion domain.

Stated simply, rule-based systems are appropriate when you possess a sufficient quantity of either: (1) practical experience in the domain, or (2) uniform case data. Rules can be derived in the first case from a logical interrogation of domain experts, and in the second case from automated pattern recognition. However, much corrosion case data are disparate, especially if obtained from multiple sources. But humans are fully capable of reasoning with disparate data, and their abilities tend to improve with increased quantity and level of detail in the available data . We often use the word "intelligence" to describe such behavior. The principal goal of this research was to determine whether we could simulate this human trait using a computer in practical engineering problems such as the prediction and diagnosis of corrosion phenomena.

Corrosion is a process, not a property, and is defined in the standards literature as the "Physicochemical interaction between a metal and its environment. . ." (ISO 1989). Generally speaking, corrosion is not a unidimensional quantity: we tend to characterize corrosion by mode as well as by rate. The properties that characterize a metal and its environment are even more numerous. Features of a metal that influence its corrosion resistance include, for example, composition, microstructure and surface finish; similarly, the corrosivity of an environment is influenced by multitudinous chemical and physical criteria such as pH, temperature, stress, etc. Finally, "corrosion resistance" and "corrosivity" are interactive variables: for example, Metal 1 might resist Environment 1 and corrode rapidly in Environment 2 while Metal 2 exhibits the reverse behavior. Because corrosion involves multiple, interacting variables a phenomenological approach to the analysis of corrosion data suggests the application of multidimensional methods.

2 Mapping Corrosion Data in a Multidimensional Feature Space

We start with a two-dimensional example of the visualization of multivariate corrosion data, the isocorrosion diagram. The abscissa typically measures the corrosive concentration, the ordinate measures the temperature, and the material is held constant at some known value. Any other factors which might influence corrosion lie somewhere outside the realm of this two-dimensional feature space, hence they are not simply unknown, they are in fact "unknowable". We expect cases of similar corrosion behavior to cluster into specific regions, i.e. relative distance measures the dissimilarity of cases. In fact, we typically draw isocorrosion lines on this diagram to delineate the regions of similar corrosion behavior.

Suppose our data consists only of the three points labeled P, O, and Q, as shown in Figure 2.

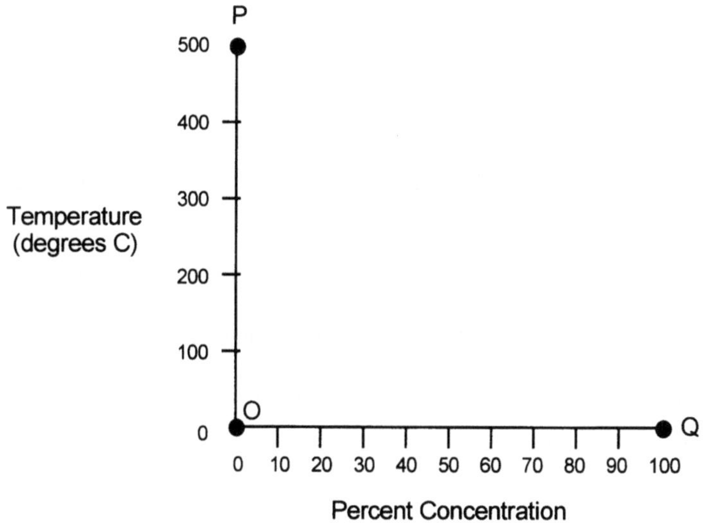

Figure 1. Three Points in 2-D Feature Space

We would expect that other data near point P would represent cases of corrosion behavior similar to that found at P, or at least more similar than that found at Q, for example. So we are comfortable with the use of distance on an isocorrosion diagram as a measure of dissimilarity.

We measure dissimilarity between cases using a weighted distance metric of the form:

$$\mathbf{d}(\mathbf{A},\mathbf{B}) = \left[\left[\sum_{i=1}^{n} \mathbf{w}_i^p \left(\left|\mathbf{A}_i - \mathbf{B}_i\right|\right)^p\right]\right]^{\frac{1}{p}} \tag{1}$$

The weights are necessary to account for difference in scale. For example, for the points in Figure 1, an unweighted metric would yield: d(O,P) = 500 degrees C and d(O,Q) = 100%. Obviously we cannot compare dimensionally different quantities. If we choose weights so as to equate each dimension's contribution to distance, or dissimilarity, and we assign the arbitrary value of 1 to the length of the diagonal PQ, we obtain the following general expression for the weights:

$$\mathbf{w}_i = \frac{\mathbf{1}}{\mathbf{N}^{\frac{1}{p}} \left(\mathbf{Max}_i - \mathbf{Min}_i\right)} \tag{2}$$

where N is the dimensionality of the feature space, which in Figure 1 equals 2. The quantity p in Equations (1) and (2) is known as the Minkowski exponent (Shepard 1980); for Euclidean space, p = 2.

As we obtain more detail about the data, we simply add more dimensions to the feature space.

3 APPLICATION TO ACTUAL CORROSION CASE HISTORIES

We assembled a database of 84 cases of corrosion failures taken from the published literature (During 1991, ASM 1979). The cases were limited to cases of the corrosion of stainless steels in primarily aqueous environments. A total of six failure mechanisms was represented in the case data (frequency in parentheses):

1. Crevice corrosion (10)
2. Flow induced corrosion (12)
3. Intergranular corrosion (9)
4. Pitting (19)
5. Stress corrosion cracking (22)
6. Uniform corrosion (12)

Our objective was to determine whether cases of the same corrosion phenomena were found to map onto singular regions of the feature space. The cases were recorded as rows in a spreadsheet in which the columns recorded specific features of the cases. The superset of features spanned 16 dimensions, and included such data as alloy composition, principal corrosive, temperature, pH, presence of chlorides and mechanical stress, etc.

To measure the dissimilarity between cases, we solved Equations (1) and (2) using the case data as input. We experimented with various values of the Minkowski exponent and achieved our best results with p = 1. The calculations were repeated for each pairing of cases, with the value of N determined *during each pairwise comparison* by counting the number of features held in common by the pair.

After finding the most similar pair, we then determined whether these two cases documented examples of the same corrosion phenomena, i.e. whether cases of identical corrosion phenomena mapped onto the same region of the feature space. We then determined the percentage of successful pairwise comparisons, comparing our results against a benchmark of 16.7% success rate, which we would expect to achieve from random guessing. Our results are illustrated in Figure 2.

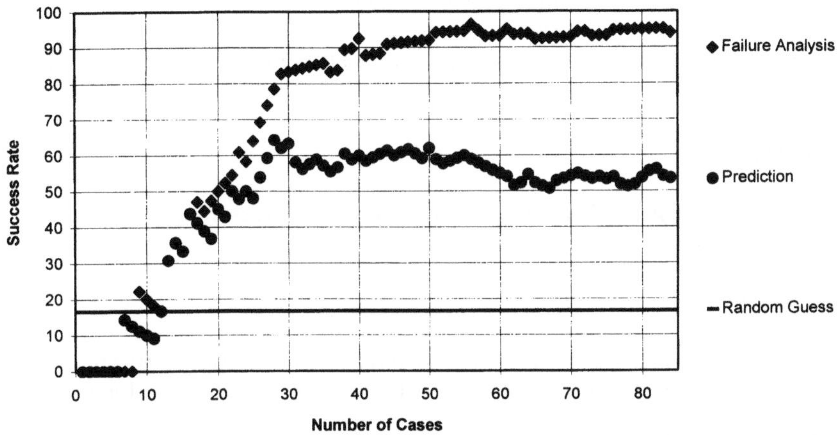

Figure 2. Number of Cases vs. Success Rate

To evaluate the system's ability to accurately predict the mode of failure, we withheld case data on macroscopic appearance and time of failure. With 12 or fewer cases known to the system, the performance was no better than the random guess benchmark of 16.7%. However, with 13 cases known to the system, the success rate jumped to 31%, and continued to rise, reaching a high value of 64.3% with 28 cases known to the system. As additional cases were introduced, the performance was degraded slightly, but never dipped below 50%, or three times better than the random guess rate.

We expected that additional detail in the case data would improve performance. To test this hypothesis we included the data on macroscopic appearance and time of failure in the calculations, to see how well the system could correctly analyze known corrosion failures. The system outperformed random guessing with only nine cases in the case base, increasing to a high value of 96.4% with 56 cases known. As before, the performance was degraded slightly as new cases were provided, but never dipped below 90%, or more than five times better than the 16.7% benchmark.

CONCLUSION

We have demonstrated a purely phenomenological approach for the prediction and analysis of corrosion failures. With a database of just 84 cases covering wide ranges of environments and corrosion phenomena, we achieved success rates for prediction and failure analysis as high as 64.3% and 96.4% respectively, which compare very favorably with the random guess benchmark of 16.7%. The methodology is applicable to any collection of phenomenological data, and requires no *a priori* knowledge of the domain. This approach avoids many of the problems associated with deriving rules based on experience, and is more appropriate for domains such as corrosion in which experts reason using prior case knowledge.

KEYWORDS

Corrosion failure analysis, knowledge-based system, case reasoning

REFERENCES

[ASM 1979] American Society for Metals, Case Histories in Failure Analysis (1979)

[During 1991] During E.E., Corrosion Atlas - A Collection of Illustrated Case Histories, Elsevier Science Publishers (1991)

[ISO 1989] ISO Standard 8044 , Corrosion of Metals and Alloys - Vocabulary. Second edition (1989)

[Shepard 1980] Shepard R.N., Multidimensional Scaling, Tree-Fitting, and Clustering, Science, Vol. 210, N° 24 (1980)

Strategic Search for Potential Quaternary Representatives of the Newest Record T_C Intermetallic Superconductor (T_C = 23K) Belonging to the New Structure Type B_2CLuNi_2

Pierre VILLARS[1], John RODGERS[2], Karin RABE[3] and Shuichi IWATA[4]

[1] *Materials Phases Data System (MPDS), Postal Box 1, 6354 Vitznau, Switzerland*
[2] *National Research Council of Canada, Ottawa KIA 0S2, Canada*
[3] *Department of Applied Physics, Yale University, New Haven, Connecticut 06520, USA*
[4] *Faculty of Engineering, University of Tokyo, 7-3-1, Hongo, Bunkyo-ku, Tokyo, 113 Japan*

Abstract

We present a global approach to the strategic search for potential quaternary high-T_c INTERMETALLICS from all potential quaternary combinations of the chemical elements (over 3.900.000 systems). It provides a list of the most promising potential quaternary high-T_c INTERMETALLICS, filtering the less probable ones via data and knowledge bases. Our predicted high-T_c superconductors are d^7 to d^{12} - La, Y, Sc, Lu, Ca, Sr - borocarbides or their boronitrides and carbonitrides.

Resume

Nous présentons une approche de la recherche stratégique de composés potentiels intermétalliques quaternaires à Tc élevé en partant des combinaisons quaternaires potentielles des éléments chimiques (plus de 3.900.000 systèmes). Elle conduit à une liste des Intermétalliques à Tc élevé les plus prometteurs, par filtrage des moins frottables à partir de bases de données et de connaissance. Les superconducteurs à Tc élevés retenus, au niveau prédictif, sont des borocarbides et leurs boronitrides ou carbonitrides des groupes d^7 to d^{12} avec les éléments -La, Y, Sc, Lu, Ca, Sr.

1 Introduction

At the beginning of this year Cava et al. [1] observed superconductivity in a multiphase B-C-Pd-Y intermetallic material with T_c = 23K. This T_c essentially ties in with the highest T_c found for an intermetallic compound, that of the thin-film $GeNb_3$ [2]. This record T_c was achieved in 1974. We now have the first progress in superconducting intermetallics in 20 years. Cava et al. [3] and Siegrist et al. [4] have explored other quaternary borocarbides and found superconductivity near 17K in.

B_2CLuNi_2 Siegrist et al. [4] have refined its crystal structure from single crystal studies and pointed out that the new superconductor crystal structure is a carbon filled-up tI10 Al_4 Ba crystal structure, which we call small atom inserted tI10 Al_4Ba type. From Pearson's Handbook [5] it is seen that this tI10 Al_4Ba crystal structure is the most populous tetragonal crystal structure, having more than 700 representatives.

2 APPLICATION OF DATABASES AND KNOWLEDGE BASES

As in the field of ternary and quaternary intermetallics there exist around 161.000 and 3.900.000 potential systems respectively, this work should provide some guidelines for selecting the most promising systems, for which experimental investigation in the search for highest-T_c intermetallic superconductors of the type B_2CLuNi_2 can first be undertaken. This search will be made by combining databases, knowledge bases and, at a later stage, inclusion of first principle calculations as well as the expertise of experimenters. The following databases are used in this work:

- Pearson's Handbook, second edition [5]. Most comprehensive, critically evaluated crystallographic database comprising around 55.000 intermetallics covering world literature (1913-1993).
- Structure Type Atlas [6]. A Pearson's Handbook companion second edition, giving for most intermetallic crystal structures 2-D and 3-D drawings as well as the atomic environments AE's (coordination polyhedra) for each atomic position (point-set).
- Ternary Alloy Phase Diagram Handbook [7]. Most comprehensive, highly standardized alloy phase diagram compendium comprising and reproducing over 15.000 ternary phase diagrams including liquid projections, isothermal sections, pseudo-binary sections and vertical sections covering world literature (1900- 1993).
- Superconductor materials database [8]. Tables 2-3 of reference [8] contain T_c of all known superconductors up to 1986. The T_c of the superconductors from 1987 to 1994 were taken from the literature.

The following knowledge bases are used in this work:

- Atomic environment knowledge base [9]. General applicable structural classification and structural relation technique based on the description of the atomic environments (coordination polyhedra) of each atom within the unit cell. This knowledge base, useful in finding structural relations (related crystal structures), supersedes the classical symmetry and size of the unit cell (number of atoms within the unit cell), as well as the sub- and super-cell relation classifications.
- Quantum structural diagram (QSD) knowledge base [10]. QSD is a 3D structure map that has proved very effective in separating intermetallics into distinct structure type domains within QSD. The coordinates are defined by atomic property expressions of pseudo-potential radii, valence electrons (group number) and electro-negativities of

the constituent chemical elements. The boundaries of such stability domains are based on separating over 7.000 intermetallics with an overall accuracy rate of 97 %. Thus QSDs can be used to predict potential intermetallics for experimentally known crystal structures by locating their structure type domain within QSD. This can be done for any potential intermetallics by calculating coordinates in QSD.

•QSD - intrinsic property relations knowledge base [11]. For high-T_C superconductors, high-T_C ferroelectrics and quasi-crystals it showed that only a limited number of the more than 3.000 experimentally known [5] crystal structures showed these properties. Furthermore, these groups of materials are narrowly located in the QSD [10]. The newest high-T_C superconductor intermetallic from Cava et al. [1] is located in the superconductor domain B of the QSD published in 1989 and is reproduced in Figure 1. The carbide superconductors are located in domain B.

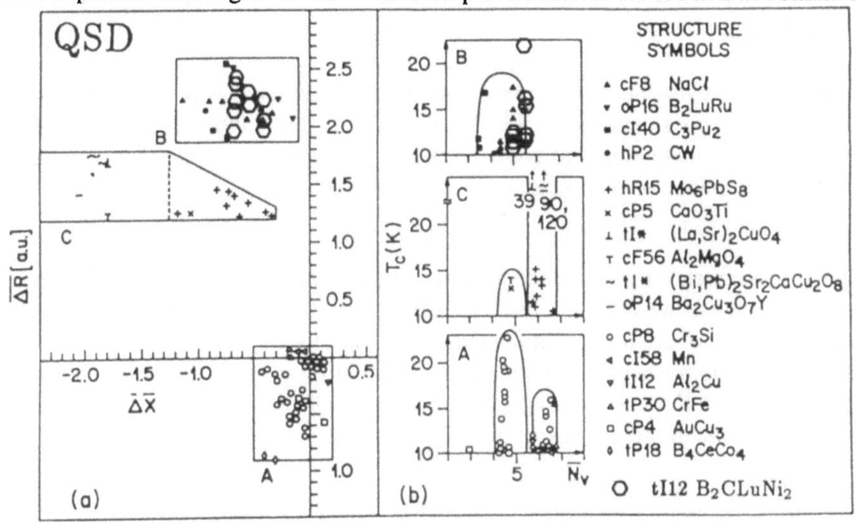

Figure 1. QSD showing all high-T_C superconductors with T_c higher than 10 K.

•Quantum formation diagram (QFD) knowledge base [12]. QFD is a 3D diagram able to separate the presently known, experimentally investigated 7000 ternary alloy (intermetallic) systems into two simple sub-spaces, the compound forming space (where at least one ternary compound is found) and the non-compound forming space (where no ternary compound is formed) with an accuracy of 95%. The coordinates for this diagram are expressions of the melting points, valence electrons (group number) and pseudo-potential radii of its constituent chemical elements.

•Atomic number (A) vs. atomic number (B) vs. atomic number (C) - materials property diagram knowledge base [13] . These plots reveal clear patterns within data for materials. The big advantage is that one has in a glance the overview of a large database in direct relation to the Periodic Table. Such patterns have been found for 'good' conductivity materials, melting temperature of materials, crystal structure of materials, as well as compound formation resp. non-compound formation within systems. In general these diagrams are used at a constant stoichiometric ratio.

- Statistical plot knowledge base [13]. Statistical plots have shown many overall regularities of data items to find relations/rules/regularities valid for a large amount of data and can thus build up a strong knowledge base. This base can be used to find 'restraints' needed for our 'filtering' purpose by examining the superconductor materials in context with the 55.000 experimentally known intermetallic compounds and phase diagrams information on more than 10.000 systems.

3 Seven Observations Using Databases and Knowledge Bases

Below we briefly list these seven observations. They are used as restraints to filter the less promising ones, starting from the more than 3.900.000 potential quaternary systems. Our ultimate aim is to come up with a list of the most promising potentially high-T_C superconductor intermetallics of the type B_2CLuNi_2

Observation 1

So far, 13 intermetallic superconductors have been experimentally found in the period from January to August 1994, with T_cs greater than 10 K and crystallizing in the tI12 B_2CLuNi_2 type. All are located in domain B of QSD (Fig. 1). The boundary of domain B in QSD was published in 1989 [11] and is based on carbide superconductors with T_c higher than 10 K. To date, all published superconductors with T_c higher than 10 K are located in one of three domains in QSD.

Observation 2

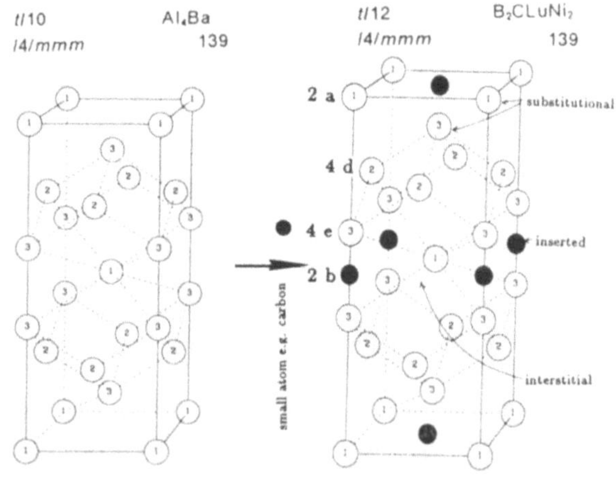

Figure 2. 3D projections of the closely related tI10 Al_4Ba and tI12 B_2CLuNi_2 structures

Between the tI12 B_2CLuNi_2 type and the tI10 Al4Ba type (with c/a- value smaller than 2.9) there exists a close structural relationship. The tI12 B_2CLuNi_2 type is a small atom-inserted tI10 Al_4Ba type. The "small atoms " occupy the inserted position with Wyckoff-notation 2(b) of space group I4/mmm (139), and is neither a substitute nor an interstitial position, but an inserted one (Fig. 2).

In the inserted position the additional atom positions itself between two "touching" atoms, in the interstitial position it positions itself in tetrahedral or octahedral holes.

Observation 3

The tI10 Al_4Ba type, depending on its c/a- value, belongs to three different coordination types (three-environment types), each with different atomic environments (coordination polyhedra), and different coordination numbers CN:

CN 8/9/22 coordination type: tI10 Al_4Ba with c/a- value lower than 2.9

CN 4/8/22 coordination type: tI10 Al_4Ba with c/a- value between 2.9 and 3.3

CN 4/8/20 coordination type: tI10 Al_4Ba with c/a- value higher than 3.3

Observation 4

The tI12 B_2CLuNi_2 belongs to the CN 6/8/9/22 coordination type and therefore is structurally closely related to the CN 8/9/22 coordination type. The following structure types belong to the CN 8/9/22 coordination type and therefore have very similar structural gross features:

tI10 $BaNiSn_3$, mC10 $CaGa_4$, tP20 $BaMg_2Sn_2$, tI12 Cu_2Sb , tI20 As_2BaPd_2, tP10 Be_2CaGe_2, mP10 Ge_2LaPt_2 and tI10 Al_4Ba with c/a- value smaller than 2.9 (in all over 800 compounds).

Observation 5

The close structural relationship between the tI10 Al_4Ba type with c/a- value lower than 2.9 and its small atom inserted tI12 B_2CLuNi_2 type has been experimentally shown in two cases. The crystal structure tI10 Al_4Ba can change to tI12 B_2CLuNi_2 just by inserting carbon, for example, (small atom) into its ternary compound crystallizing in the tI10 Al_4Ba type:

B_2Co_2Y (tI10 Al_4Ba, published in 1973 [5]) , B_2CCo_2Y (tI12 B_2CLuNi_2, published in 1994 [8])
B_2Co_2La (tI10 Al_4Ba, published in 1973 [5]),B2CCo2La (tI12 B_2CLuNi_2, published in 1994 [8])

Observation 6

About 780 experimentally known ternary intermetallics crystallizing in one of the above listed eight crystal structures belonging to the CN 8/9/22 coordination type show the following chemical element combinations: XY_2Z_2, where X is either a f- or s^2N element (occupying the position with Wyckoff notation 2(a) of space group

I4/mmm) , Y is a d^6- to d^{12}- element (occupying position 4(d)), and Z is a p^3- to p^5-element (occupying position 4(e)) .

Observation 7

Within the tI12 B_2CLuNi_2 type there exists a geometrical restraint which requires that two of the four chemical elements have to be small sized atoms, the ones occupying positions with Wyckoff notation 2(b) and 4(e) of space group I4/mmm. The 2(b) position represents the location of the small atom inserted position and the 4(e) position is the p-element position of the tI10 Al_4Ba type, space group I4/mmm (139). This geometrical restraint is best seen by looking at the atomic environments of the positions 4(e) of both structure types (Fig. 3).

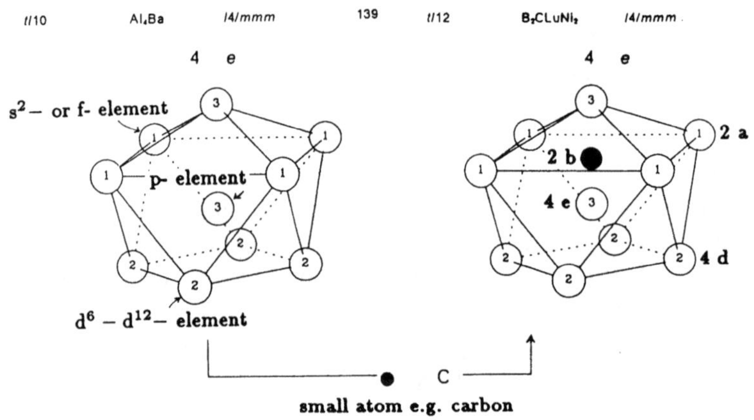

Figure 3. Atomic environments of the 4e position (I4/mmm) for the closely related tI10 Al_4Ba and tI12 B_2CLuNi_2 types

In the tI10 Al_4Ba type the p-element is surrounded by nine neighbor atoms (the shortest distance is about 0.26 nm between the p-element atoms), in the tI12 B_2CLuNi_2 type the p-element forms with the inserted small atom a dumbbell and this dumbbell is surrounded by the same nine neighbor atoms. The distance between the p-element of the 4(e) position and the inserted small atom of the 2(b) position forming the dumbbell is about 0.15 nm , which represents the center of the nine neighbor atoms. Such dumbbells can only be formed by fully directional bonding; therefore it is sensible to use covalent radii for geometrical considerations. Below we list the covalent radii for all "small atoms":

p- elements: r(B)= 0.085nm, r(C)= 0.077nm, r(N)= 0.070nm , r(O)= 0.066nm, r(F)= 0.07nm, r(Al)= 0.126nm, r(Si)= 0.12nm, r(P)= 0.11nm, r(S)= 0.10nm
s- elements: r(H)= 0.028nm, r(Be)= 0.10nm, r(Li)= 0.13nm, r(Mg)= 0.14nm, r(Na)= 0.15nm

Therefore the only potential elements are B, C, N, O, F, H, Be, for the p-element position 4(e) B, C, N, O or F, for the inserted small atom position 2(b) B, C, N, O, F, H or Be.

4 SCREENING PROCEDURE

Based on the seven observations mentioned we present a list of the most promising potential high-T_C superconductor intermetallics of the type B_2CLuNi_2 by:

1. Generating all possible XY_2Z_2 combinations, where X= f- or s^2 -element, Y= d^6- to d^{12}- element and Z= B, C, N, O or F. This step is based on observations 2, 3, 5, 6,7.

2. Filtering those potential compounds in QSD located in domain B. This step is based on observations 1, 2 and 5.

3. Screening the potential compounds located in domain B in the QFD from the non-compound forming systems. This step is carried out to ensure eliminating systems which are most likely non-compound forming systems.

4. Highlighting the potential compounds for which ternary compounds have already been experimentally found crystallizing in the CN 8/9/22 coordination type; as for those compounds the uncertainty of its ternary starting compound is eliminated. This step is based on observation 5.

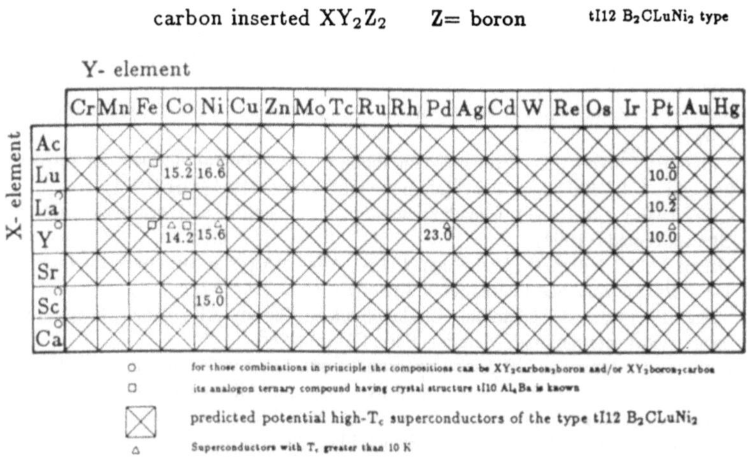

Table 1. Predicted quaternary high- T_c superconductor intermetallics of the type tI12 B_2CLuNi_2

Table 1 shows the predicted most potential high-T_C superconductors of the type B_2CLuNi_2. T_c and Ac containing compounds of possibly less interest. Those compounds which have already been found to be superconductors are represented by their T_c in K. To sum up, the d^7 to d^{12} - La, Y, Sc, Lu, Ca, Sr- borocarbides are the most potentially high - T_c superconductors of the type B_2CLuNi_2. These borocarbides could also be boronitrides or carbonitrides. The boronitides and carbonitrides are less likely because no ternary carbide or nitride has been experimentally found among the 780 compounds belonging to the coordination type CN 8/9/22. It is important to realize that our predicted superconductors meet all conditions set by the 12 new intermetallic superconductors found experimentally to date. The condition set by QSD is necessary but not sufficient, which means that the probability of intermetallics in domain B being a superconductor is much greater. Nevertheless intermetallics located outside domain B in QSD have until now always been shown not to be superconductors with T_c higher than 10 K.

Keywords

high-T_C Superconductor intermetallics, quaternary element combination, B_2CLuNi_2

References

[1] Cava R.J., Takagi,H., Batlogg B., Zandbergen H.W., Krajewski J.J., Peck Jr. W.F., van Dover R.B., Felder R.J., Siegrist T., Mizuhashi K., Lee J.O., Eisaki H., Carter S.A., Uchida, S., Nature 367 146-148 (1994)

[2] Fisk,Z., Nature 367 117-118 (1994)

[3] Cava R.G., Takagi H., Zandbergen H.W., Krajewski J.J., Peck Jr. W.F., Siegrist T., Batlogg B., van Dover R.B, Felder R.J., Mizuhashi K., Lee J.O., Eisaki H., Uchida S.,Nature 367, 252-253 (1994)

[4] Siegrist T., Zandbergen H.W., Cava R.J., Krajewski J.J., Peck Jr. W.F., Nature 367, 254-256 (1994)

[5a] Villars P., Calvert L.D., Pearson's Handbook of Crystallographic Data for Intermetallic Phases, 2nd edn., 1-4 5'300 pages, ASM International, Materials Park, OH (1991)

[5b] CRYSTMET CAN/SND, Canadian Institute for Scientific and Technical Information, NRCC, Ottawa, Canada (1994)

[6] Daams J.C.L., Villars P., Van Vucht J.H.N., Atlas of Crystal Structures for Intermetallic Phases, 1-4 7'300 pages, ASM International, Materials Park, OH (1991)

[7] Villars P., Prince A., Okamoto H., Handbook of Ternary Alloy Phase Diagrams, 1-10 14' 000 pages, ASM International, Materials Park, OH (in print).

[8] Phillips J.C., Physics of High-T_C Superconductors, 336-374, Academic Press, New York (1989)

[9] Daams J.L.C., Villars P., Van Vucht J.H.N., Journal of Alloys and Compounds, 182 1-32 (1993)

[10] Villars, P., Hulliger F., Journal of the Less-Common Metals 182 289-315 (1987)

[11] Rabe K.M., Phillips J.C., Villars P., Brown I.D., Physical Review B, 45B 7650-7676 (1992)

[12] Villars P., Journal of the Less-Common Metals 119, 175-188 (1986)

[13] Rodgers J., Villars P., Journal of Alloys and Compounds 197 167-176 (1993)

Chapter 3

MODELING TOOLS FOR KNOWLEDGE DISCOVERY

PATTERN RECOGNITION AND CLASSIFICATION IN MULTIVARIATE PROBLEMS: MORPHOLOGY FOR THE DETERMINATION OF USEFUL BOUNDARIES. INDUSTRIAL APPLICATIONS

Marie-Françoise DOUTRE

Consultant expert, 101 rue Chardon Lagache, 75016 Paris, France

ABSTRACT

In numerous fields, it is essential to classify behavior, results and representation globally, while the underlying phenomena depend on many parameters. Experience and knowledge of these phenomena are needed for good classification. We propose an original and efficient three step method for overcoming difficulties: 1, characterizing phenomena through existing data, 2, creating a two-dimensional image, 3, searching shape and its useful "contour". This method permits prediction of new data from existing data. It is very successful in multivariate quality control, anticipating failure, optimizing maintenance and classifying behavior and images .

RESUME

Dans de nombreux domaines il est important de classer des comportements, des résultats, des représentations, sous une forme globale, alors que le "phénomène" sous-jacent dépend d'un grand nombre de paramètres, et que seule l'expérience et la connaissance du phénomène permettent ce classement. La méthode décrite ci-après offre une solution originale et efficace. La méthode se décompose en trois phases : caractérisation du phénomène à partir des données existantes, réalisation d'une image en deux dimensions, et recherche de la forme et du contour utile. Cette méthode excelle dans le contrôle qualité multivarié, la prévision des pannes, l'optimisation de la maintenance, la classification des comportements et des images.

1 INTRODUCTION : PROBLEMATICS

Let us suppose a quality control of an automobile opening : this implies comparing more than a hundred measurements for one given datum, e.g. water-tightness, to theoretical nominal values.

Let us now suppose a manufactured cheese whose physical characteristics are perfectly known, and its appreciation by a customer panel : this implies combining numerous physical characteristics with subjective qualitative variables. Then let us suppose characterizing a network or complex materials by given specifications, synthesizing a large number of quantitative and qualitative parameters in an often fuzzy environment. In these examples (and many other fields have identical problems), one must diagnose, select, optimize a phenomenon depending, first, on catalogued but numerous parameters with often unknown interactions and, second, unknown or hard-to-evaluate parameters. The global decision-making criterion (quality, customer satisfaction, specification adequacy) parameters and their interaction must be considered. This criterion is linked to experience and expertise; this supposes considering the past and, thereby, experience or experimental data.

The article thus turns on three approach axes and one application axis. Graphics are simulated, but they all have concrete reality.

1 Determining the "pattern" characterizing the phenomenon
2 Determining the "image" of the phenomenon
3 Morphology to determine useful pattern and boundary
4 Industrial applications.

2 DETERMINING THE "PATTERN" CHARACTERIZING THE PHENOMENON OR " ON THE IMPORTANCE OF CHOOSING DATA PREPARATION"

Definitions :

We call "characteristic vector" of a phenomenon the vector of p dimension composed of the parameters "a priori" retained to characterize the phenomenon.

We call "characteristic pattern" the set of nxp values taken by the p parameters for the n measurements (also called "realizations" or "points"). We can therefore consider either the matrix $A = (a_i^j)$ of the shape, with a_i^j = value of the jth parameter for the ith point-measurement, or the whole set of characteristic n vectors : $V_i = (a_i^j)_{j=1 \text{à} p}$, or yet the set of p vector-parameters Ê: $P^j = (a_i^j)_{i=1 \text{à} n}$

The choice of n initial realizations corresponds to the choice of the learning set.

Notations :

$$\text{average : } \bar{a}^j = \frac{1}{n}\sum_{i=1}^{n} a_i^j$$

$$\text{standard deviation : } (\sigma^j)^2 = \frac{1}{n}\sum_{i=1}^{n}(a_i^j - \bar{a}^j)^2$$

$$\text{correlation coefficient : } \rho_{hk} = \frac{\sum_j (a_i^h - \bar{a}^h)(a_i^k - \bar{a}^k)}{n.\sigma^h \sigma^k}$$

principal direction : $A^t.A$'s eigen vector.

The initial step of the study consists in choosing these parameters and the learning set, i.e. transforming a problem into a set of data expressing this problem.

Figure 1. The data of the problem

2.1 Parameter relevance

Two relevance levels must be considered : do we have "characteristic" parameters (in sufficient number and quality), and/or are some parameters redundant? This leads to two types of search.

•.Search for representativity (by working groups using techniques of problem solving and decisional diagrams, or else by coherence studies of the graphic displays when there are indications as to the parameter laws)

• Search for redundant parameters through correlation search (by statistical calculations) :

Ph and Pk correlated $\rho_{hk} \geq 75$ % (this is the usually admitted value)
or by concomitant variations corresponding to the cause-effect couple
or by proximity in a principal plane (cf. 2-1)
$[d(Ph, Pk)]^2 = 2 (1 - \rho_{hk}))$

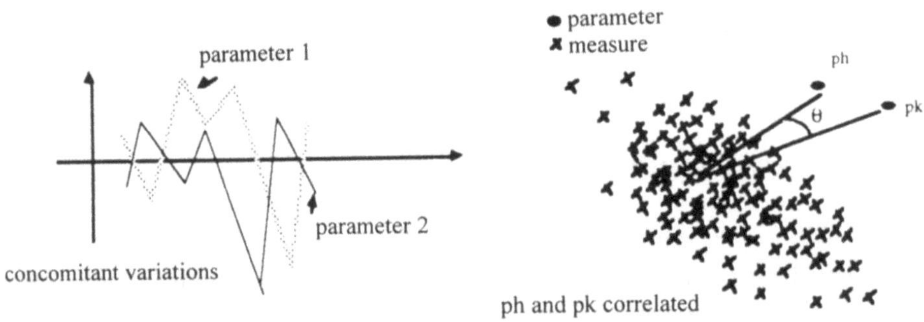

Figure 2. Redundant parameters.

2.2 Search for "Homogeneous Zones"

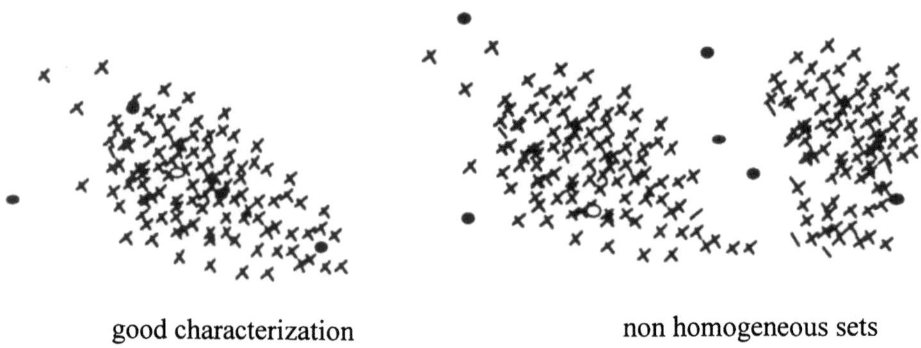

Figure 3. Homogeneity

To be efficient the method must be applied to a set of points defined by parameters "precisely" expressing the problem. There must therefore be "characterization" of this problem on the level of the parameters and of the learning set.

Homogeneity is approached by visualizing the pattern in a space and observing the number of the clusters obtained, e.g. in dimension 2 :

One can also use a technique of dynamic clusters and of agglomeration around kernels, then count the clusters obtained : one selects the initial kernel and assigns each point to the "class" of kernel to which it is closest, according to the chosen metrics.

These techniques are limited, the first by the choice of the projection space, the second by the choice of initial kernels. When the data are not homogeneous, one must either refine the initial data, or reexamine the parameters and separate the set of data so as to define sub-problems.

with 2 kernels with 3 kernels

Figure 4. Clusters with dynamic clouds

2.3 Eliminating spurious points

Another preliminary task must be carried out : the study of data reliability. Very often certain errors creep into a set of data, errors due to human behavior or to a material dysfunction. Two techniques allow us to eliminate most of these errors : the first involves calculating the average and the typical divergence in the learning set, for each parameter, and then noting the points that are further than two or three from the average $\bar{a}_i^j$. When using this method, it must not be forgotten that these divergent points were considered in calculating the statistics and that they therefore perturb this selection; it is thus essential to carry out several repetitions after each refinement.

$$(\rho^j)^2 = \frac{1}{n}\sum_{i=1,\neq i0}^{n}(a_i^j - \bar{a}_i^j) + (a_{i0}^j - \bar{a}_i^j)^2 \text{ with } (a_{i0}^j - \bar{a}_i^j)^2 >> (a_{i0}^j - \bar{a}_i^j)^2$$

The second technique is graphic. When the data are projected onto a plane, the divergent points "absorb" a large part of the information and behave like elastics, and can thus be easily seen.

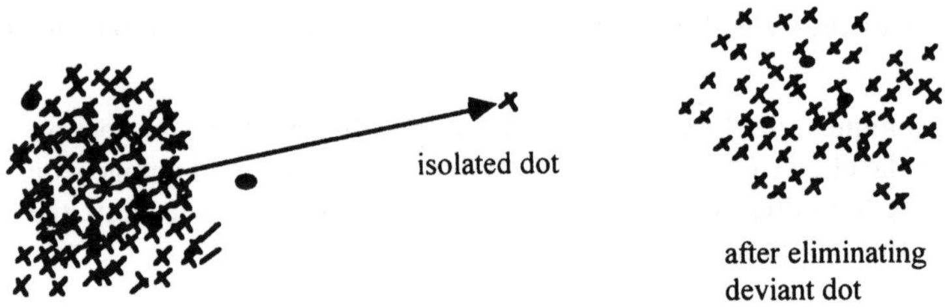

Figure 5. Spurious point

Thus, for a principal plane :

$$u^t.A^t.A.u = u^t.A^{*t}.A^*.u \ + \ u^t.Vi_0{}^t.Vi_0 \ .u$$

$$\text{avec } u^t.Vi_0{}^t.Vi_0{}^t.u \ >> u^t.A^{*t}.A^*.u \ \text{ et } \ A^* \ =(a_i^j).i \neq i0$$

3 DETERMINING THE "IMAGE"

At this point, the characteristic pattern is supposed indeed "characteristic", homogeneous and without spurious points; In this part we shall see how to achieve a two dimensional display of this pattern : the image of the pattern (possibly several images in different planes).

3.1 PCA or correspondence analysis

A first method involves searching for the plane that provides the most information and, in general, (although in some cases one can work directly on raw data) one normalizes the data.

a_i^j is replaced by $:\alpha_i^j = \dfrac{a_i^j - \overline{a_i^j}}{\sqrt{n\sigma^j}}$ the matrix A becomes the matrix Λ

Then one minimizes the sum of the square of the distances of the point-vectors V_i with regard to a plane, which means finding u so that ut. Λ^t. Λ.u is maximal.

$u^t.\Lambda^t.\Lambda.u$ maximal $\Leftrightarrow$ u, $\Lambda^t.\Lambda$'s eigen vector associated with λ's eigen value.

The "amount of information" for a dimension space q composed of q eigen vectors is given by : $\frac{\sum_{i=i\sigma n}\lambda_i}{\sum_{i=i\sigma n}\lambda_i}$

For a properly conducted preliminary study, one generally obtains principal planes corresponding to amounts of information greater than 70% and to following classical patterns : (due to the "normality" of the parameters).

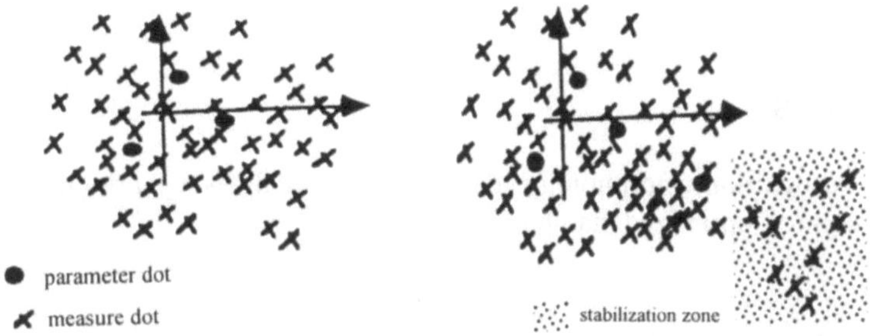

Figure 6. Image by PCA

3.2 Derived methods

In the analysis method using principal component parts, one distinguishes parameters and measurements and selects classical Euclidian metrics. It is nonetheless possible to use analogous methods with different metrics (distance from the chi -2), making it possible to have an identical role played by the lines and columns of the initial matrix. Correspondence analysis thus provides excellent results.

3.3 Other methods

Naturally, all projection based methods can be used, but the basic precaution to be taken is that of knowing the representativity of the projection with regard to the initial data.

One can thus use dynamic clusters, the variance analysis method or modeling by design of experiment.

CONCLUSION

This method of analyzing data, then searching for morphological, useful or theoretical shapes or contours, makes it possible to solve the problem of not knowing the interactions of numerous parameters on a phenomenon, as well as that of not considering certain influential but misunderstood or hard to measure influential parameters. It can be applied in many fields and helps to solve many problems. It is based on reality, since the initial data, called learning, allow us to determine the limits, and yet this reality can be considered gradually, can be modified, either theoretically or interactively during the development of the phenomenon or of the process.

This method is based on statistics and topology and is, therefore, complex, but its application requires only good experience and good expertise in the field studied.

KEYWORDS

Pattern recognition, morphology, data analysis, analysis by principal components, quality control, optimized maintenance, indicators.

INDEX OF GRAPHICS :

REFERENCES

[1] Lebart L., Morineau A.,Traitement des données statistiques,Dunod-Bordas (1977)
[2] Diday E., Lemaire J., Eléments d'analyse de données ,Dunod (1982)
[3] Fukunaga K., Introduction to statistical recognition, Academic Press (1972)
[4] Dubuisson B., Diagnostic et reconnaissance de formes, Herm s (1990)
[5] Serra J.,Image analysis and mathematical morphology, Academic Press (1982)
[6] Haralick R., Sternberg , Image analysis using mathematical morphology IEEE PAM-9 (1987)
[7] Golay M., Hexagonal parallel transformations IEEE C-18 (1969).
[8] Doutre M.F., Thèse UTC Compiègne (1991).

Maximum Likelihood Processing of Antenna Array Signals

André MARGUINAUD

Alcatel Espace, 5 rue Noël Pons, 92737 Nanterre cedex, France

Abstract

The antenna array considered here is composed of m receiving elements whose output signals are analytically digitized. Covariance matrix estimator and the correlation indices of its estimated entries are derived from the Maximum Likelihood Principle (ML). Practical methods for direction finding and Jammer cancellation can be designed to approach the optimum deduced from ML. Implementation complexity is reduced by suitable choice of dimensioning of data representation during processing.

Resume

L'antenne considérée est composée de m éléments récepteurs dont les signaux sont numérisés analytiquement. L'estimateur de la matrice de covariance et les indices de covariance des paramètres estimés sont élaborés à partir du principe du maximum de vraisemblance (ML). On peut développer des méthodes pratiques de goniométrie et d'annulation de brouilleur qui s'approchent de l'optimum ML. La complexité de mise en oeuvre est réduite par un choix convenable du nombre de dimensions au cours des calculs.

1 Introduction

The antenna array considered here is composed of m receiving elements, whose output signals are analytically digitized before beam-forming. Sampled representation of band-limited signals cannot be justified by the existence of an explicit interpolation formula. However, combinations of suitable hypothesis testing and linear interpolation can do so.

This sampling principle, along with a locally uniform representation of continuous signals, allows treatment of the actual signals as weighted sums of sinusoids within the physical band.

This fundamental property provides a practical method to calculate the analytical sampling from the real sampling in a much simpler way than can be realized by a digitalization of the Hilbert transform function.

Maximum Likelihood estimation of single jammer parameters shows that these parameters are solutions of a linear system of equations defined by the covariance matrix: one Eigen-value determines the jammer power, the associated Eigen-vector characterizes the direction of propagation of the jammer signal. Unfortunately in the case of multiple radio-frequency sources, whereas the resulting covariance matrix is the sum of the elementary covariance matrices, this linear property is not verified by the Eigen-vectors. However the difference of two covariance matrices taken at closely approached instants in time is a covariance matrix given by a simple explicit expression, so that optimum and actual performance levels can be quantitatively evaluated (*direction and ranging accuracy, jammer cancellation, etc ...*).

Complexity, which is a linear function of the number of elementary antennas m and of jammers is obtained by taking the dimension adapted to the problem: $m(m+1)/2$ for covariance matrix estimation, m for jammer cancellation, 2 for geographical location, etc ...

The proofs and detailed calculations concerning the results presented in this paper have been given in **[1]** so we shall emphasize the motivations and philosophy of the solutions described. An antenna array is a convenient receiving system when mechanical precision is hard to achieve. e.g. deep space antennas, large antennas deployed from a satellite (Fig. 1).

The following results have been obtained during a DRET contract [4].

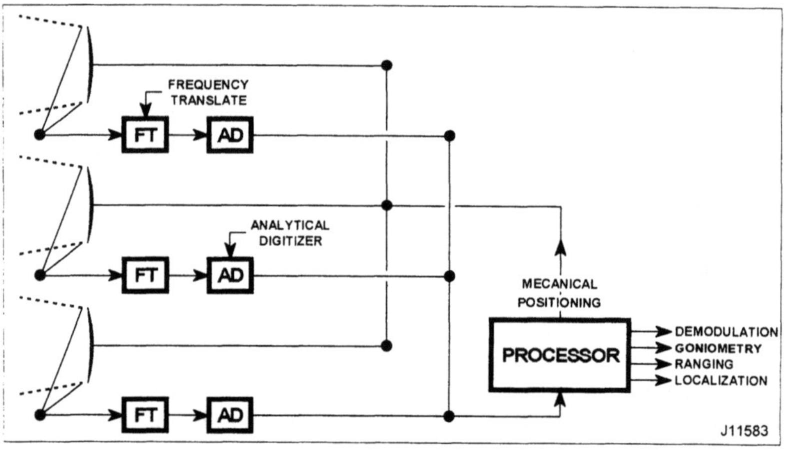

Figure 1. Analytic array

2 Analytical Sampling

Sampled analytical representation of the received signal by each element of the antenna array is well adapted to digital processing. Some people, however, fear that some performance degradation may result from this "**reduction**" of continuous band limited functions to the equivalent of entire series. Their caution stems from the unsound foundation of the (*analytic*) sampling principle.

For complex functions with a continuous Fourier transform, the famous Shannon interpolating formula (1) can be proven. However this formula diverges for any sinusoid even when its frequency lies inside the bandwidth/passeband.

(1)
$$\begin{cases} z(t) = \sum_{k=-\infty}^{+\infty} z\left(\frac{k}{\Delta B}\right) \frac{\sin \pi(t\Delta B - k)}{\pi(t\Delta B - k)} \\ \exp 2\pi j f t \neq \sum_{k=-\infty}^{+\infty} \frac{\sin \pi(t\Delta B - k)}{\pi(t\Delta B - k)} \exp 2\pi j \frac{kf}{\Delta B} \end{cases} \quad \text{is divergent}$$

ΔB is the total one sided bandwidth of the signal.

This divergence can be removed by CESARO summation as proposed by FEJER. The time sampling principle is well established, but one should be careful with interpolation formulae.

For practical purposes in signal processing, it is convenient to complete this principle by a **local sine decomposition principle** which means that any time continuous and bandlimited signal can be locally considered as a finite sum of sines the frequencies of which lie in the physical band. Most algorithms rest on the application of this principle, without explicitly calculating these components. A pertinent use of this procedure is the determination of the weights used to calculate analytical sampling from real sampling.

For a given signal to noise quadrature ratio ρ, we obtain a simpler solution than by digitizing the Hilbert transform [cf. (2)].

(2)
$$\begin{cases} x(t) \underline{\underline{\Delta}} r\left(t - \frac{1}{2Fs}\right) + r\left(t + \frac{1}{2Fs}\right) - r\left(t - \frac{3}{2Fs}\right) - r\left(t + \frac{3}{2Fs}\right) \\ y(t) \underline{\underline{\Delta}} r\left(t - \frac{1}{2Fs}\right) - r\left(t + \frac{1}{2Fs}\right) + r\left(t - \frac{3}{2Fs}\right) - r\left(t + \frac{3}{2Fs}\right) \end{cases}$$

$$\begin{cases} Z(t) \underline{\underline{\Delta}} x(t) + jy(t) \qquad 2\pi F \frac{1}{Fs} = \pm \frac{\pi}{2} \bmod 2\pi \\ \rho(F + \Delta F) = 1 / \left(tg\pi \frac{\Delta F}{Fs}\right)^6 \quad \textit{independent of the central frequency } F. \end{cases}$$

Remark: The proof of the local sine decomposition derives from the following theoretical results :

- local WEIRSTRASS uniform approximation of real continuous functions by sines,
- quadrature of cos(ωt+φ) is sin(ωt+φ),
- time sampling principle,
- frequency equivalence modulo Fs due to sampling at frequency Fs.

3 COVARIANCE MATRIX PRESENTATION

High resolution localization techniques (cf. **[2]**) are characterized by the fact that the resolution power is limited neither by beam-width nor by signal to noise ratio, but only by the integration time and the propagation model knowledge.

The geometry (Fig. 2) and the linearity behavior of radiating elements shows that the phase difference between any two radiating elements depends only on the direction of signal arrival even in the case of mutual coupling

Figure 2. Analytical digitizer

	$t-3\Delta t$	$t-\Delta t$	$t+\frac{1}{Fs}$	$t+\frac{3}{Fs}$
x (t) SYMMETRIC WEIGHING	-1	3	3	-1
y (t) ANTISYMMETRIC WEIGHING	-1	-3	3	1

ΔF HALF BANDWIDTH

$Fs = \frac{1}{\Delta t}$ SAMPLING FREQUENCY $\qquad \rho = \frac{1}{\left(tg\,\pi \frac{\Delta F}{Fs}\right)^6}$

J11584

The coefficients of the estimated covariance matrix as defined by (3) where $\hat{z}_\ell\,(t)$ represents the analytical sample obtained for element l at the time t, are estimations of these direction characteristics.

$$(3) \qquad C_{\ell k} \underline{\Delta} \frac{1}{T\Delta B} \sum_t \hat{z}_\ell\,(t)\,\hat{z}_k^*(t) \Rightarrow C_{\ell k} = C_{k\ell}^*$$

The scaling factor TΔB is the product of the integration time multiplied by the one sided noise bandwidth ΔB equal to the analytic sampling rate so that TΔB is the number of samples.

In the case of d electromagnetic source directions, the analytic sample takes the following form, where $n_\ell(t)$ represents the complex additive noise at element ℓ.

$$(4)\quad \begin{cases} \hat{z}_\ell(t)=\sum_{i=1}^{d} z_{\ell i}(t)+n_\ell(t), \\ <n_\ell(t)>=0, <n_\ell(t)n_\ell(\tau)>=\sigma^2\delta_{t\tau}, <n_\ell(t)n_k(t)>=\sigma^2\delta_{lk} \end{cases}$$

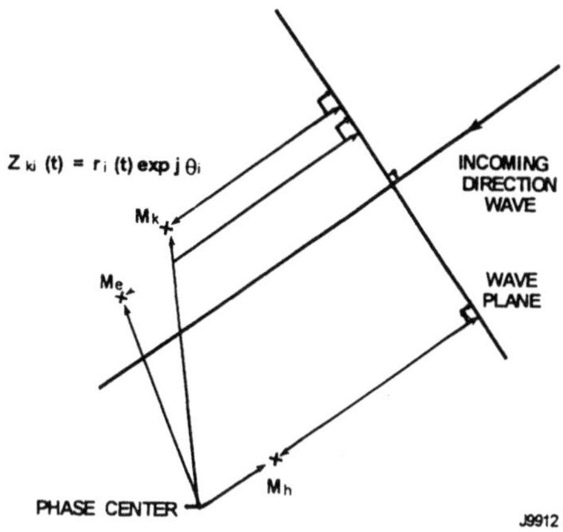

Figure 3. Array geometry principle

For a constant amplitude signal, one can write:

$z_{\ell i}(t)\overset{*}{z}_{ki}(t)=r_i^2 \exp j(\theta_{\ell i}-\theta_{ki}), with\, r_i^2$ is practically independent of the radiating elements.

Knowing the $\frac{m(m+1)}{2}$ values $C_{\ell k}$ defined by (3), the estimation of the d pairs of real values $(r_i^2, \theta_{\ell i}-\theta_{ki})$ by the Maximum Likelihood principle is nearly equivalent to find the solution which minimizes the sum of the m calculated noise variance $\sum_t n_p(t)\, \overset{*}{n}_\ell(t)$..

In the case of a single source the Eigen-value of the covariance matrix is proportional to the source power and its corresponding Eigen-vector is the m dimensional vector $(\exp j\theta_l)$.

Unfortunately, this illuminating interpretation cannot be extended to the general case of d decorrelated sources, since the global covariance matrix practically equals the sum of the d elementary covariance matrices as shown by (5).

$$(5)\qquad <C_{\ell k}>=\sum_{i=1}^{d} r_i^2 \exp j(\theta_{\ell i}-\theta_{ki})+\sigma^2\delta_{\ell k}$$

Only diagonal terms have a noise contribution. The covariance coefficients calculated by (3) are correlated and the fundamental relation (6) obtained after some lengthy calculations gives all the centered correlation indices.

$$(6)\quad \begin{cases} <C_{\ell k}C_{qr}>-<C_{\ell k}><C_{qr}>=\dfrac{1}{T\Delta B}<C_{\ell r}><C_{qk}> \\[2ex] <C_{\ell k}C^{*}_{\ell k}>-<C_{\ell k}><C^{*}_{\ell k}>=\dfrac{1}{T\Delta B}<C_{\ell\ell}><C_{kk}>=\dfrac{\left(\sigma^2+\sum_i r_i^2\right)^2}{T\Delta B} \end{cases}$$

Geometrically the first relation (6) is easily described as shown in Figure 4, the second relation is obtained by specializing the first : the rectangle becomes a square. This correlation can be lowered to any level by increasing the number TΔB of samples.

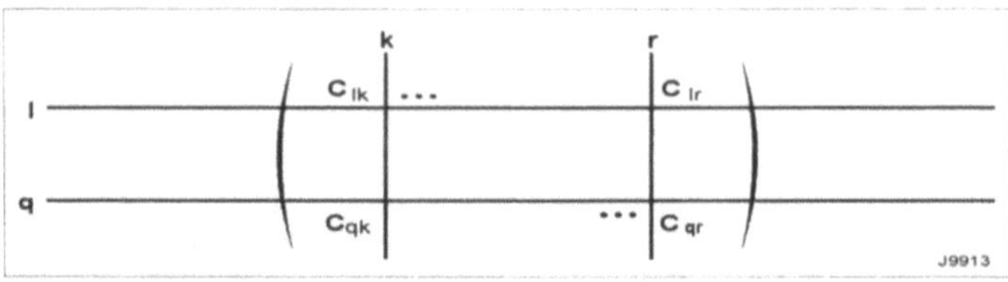

Figure 4. Centered correlation between correlation matrix coefficients

4 EXPLOITATION OF THE ESTIMATED MATRIX

Standard methods like MUSIC and ESPRIT [3] consider that a direction i is defined by an m vector the components of which are of the form $r_{\ell i} \exp j\theta_{\ell i}$.In this last expression, the main parameter is the angle $\theta_{\ell i}$ which is defined modulo 2π up to a common value to the m radiating elements.

In the case of a single radiating external source (5) we suggest using relation (7) below to estimate the m components of the direction vector

$$(7)\quad r\ \exp j\theta_\ell = \left(\prod_{k=1}^{m} C_{lk}\right)^{\frac{1}{m}}$$

For more than one direction, the open technical literature relies on the extraction of Eigen-values and Eigen-vectors of the estimated covariance matrix. The implementation complexity and the existence of numerical instability in the case of Eigen-value equality encourages us to try another method. This method which is described below rests on simple principles.

- an external source direction is nearly constant in time,

- the difference of two covariance matrices (CM) taken at closest approach instants in time is a covariance matrix which in general corresponds to either zero or a single radiofrequency direction,
- to test that a CM corresponds to zero source, one compares its trace to a predetermined level,
- to test that a CM corresponds to a single source, one compares the modules of its $\frac{m(m-1)}{2}$ non diagonal entries to the mean of its diagonal elements,
- the activity level of an external source is proportional to the modulus of the scalar product of the m signal vector $z_\ell(t)$ by the direction vector.

Using these ideas jammer cancelling and goniometry can be performed from single source covariance matrices which leads to a complexity level which is linear in the number of radiating elements and in the external radiating sources.

Jammer cancellation is carried out in an m-dimensional complex space by estimating the jammer contribution on each radiating element, then subtracting this estimation and finally summing up the m resulting complex samples in the useful direction (m dimensions) to get a complex estimation of this so called useful signal.

The efficiency of this cancellation is measured by the increase of the signal to noise ratio from :

$\rho = \frac{S^2}{J^2 + \sigma^2}$ *to* $\rho = \frac{S^2}{\sigma^2}(1 - \Delta G \Delta G^*)$, where the discriminant coefficient ΔG is given by (8).

(8) $$\Delta G \underset{=}{\Delta} \frac{1}{m} \sum_{i=1}^{m} exp\, j(\theta_{\ell j} - \theta_{\ell s})$$

As expected, the final ρ equals zero when the signal and the jammer are in the same direction since $\theta_{\ell j} = \theta_{\ell s}$.

To obtain a closed form for the variance $<\left(\frac{\Delta D}{D}\right)^2>$ of the source direction D of arrival, mutual coupling effects have been neglected. The new parameters appearing in formula (9) have the following significance : λ is the carrier wave length and R the gyration radius of element nodes.

(9) $$<\left(\frac{\Delta D}{D}\right)^2> = \frac{2\lambda^2}{m^2 R^2 T \Delta B}\left(1 + \frac{\sigma^2}{r^2}\right)^2 > \frac{2\lambda^2}{m^2 R^2 T \Delta B}$$

This formula proves the basic result that angular measurement is neither limited by the signal to noise ratio nor by beam-width. Expression (9) can be used for operational consideration.

In the practical situation, mutual coupling cannot be neglected, and the problem is to convert m dimensional vectors indicating d direction to 2 dimensional vectors, the components of which are ordinary angles. This conversion is best realized by interpolation from geographically known radio-frequency sources.

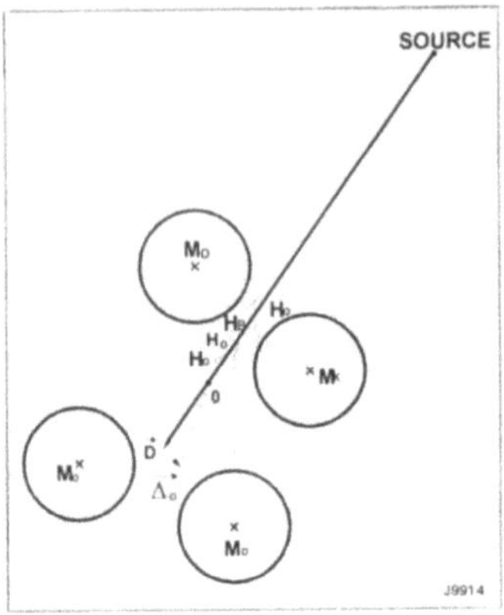

Figure 5. Source goniometry

CONCLUSION

A general method has been proposed to process received signals from elements of an antenna array, keeping the complexity to an acceptable level (linear with m and d), and giving the main operational performance. The conceptual tools are least-square estimation and hypothesis testing. Catastrophic complexity expansion has been avoided by suitable choice of dimensioning of data representation during processing.

KEYWORDS

Maximum likelihood, Analytical sampling, Antenna Array, Covariance matrix, Source direction estimation, Jammer cancellation.

REFERENCES

[1] Marguinaud A, Réseau d'antennes analytiques, Tremplin Technologique ESIEE (1994)
[2] Bienvenu G., Kopp L., Méthodes haute résolution pour la localisation des sources rayonnantes, Onde électrique - Juillet - Août (1984)
[3] Roy R.H., Esprit - Estimation of signal parameters via rotational invariance techniques, PhD dissertation, Stanford University (1987)
[4] Contrat DRET N°92-373 , Sous-système de localisation.

Revisiting the Electrical Effects of Substituents: Challenging the Basic Substituent Scales

Gérard THIROT, Aliette COSSE-BARBI and Jacques-Emile DUBOIS

Institut de Topologie et de Dynamique des Systèmes (ITODYS), Université Paris VII, 1, rue Guy de la Brosse, 75005 Paris, France

Abstract

Modeling substituent effects by the fairly successful models of YUKAWA-TSUNO (YT) and CHARTON (MC), one has recourse to several standardized scales that hide incorrect implicit hypotheses. The integration of various correlations to obtain a "unique scale" of substituent parameters has led to a confused situation. We propose an ICAD (I = Inductive, C = Conjugative, A = Accepting, D = Donating) equation based on the statistical identification analysis of four structural group factors involved in setting up a quadruple similarity scale (ICAD) with one inductive and three-resonance contributions. We have shown that our ICAD scale can more successfully handle numerous physical and reactional situations, than complex LFER'S.

Resume

Dans la modélisation des effets de substituants Xi dans le cadre des modèles bien acceptés de YUKAWA-TSUNO (YT) et CHARTON (MC), on utilise diverses échelles standardisées développées à partir d'hypothèses implicites en partie incorrectes. L'intégration de corrélations diverses pour les fondre dans une "échelle unique" de paramètres de substituants a conduit à une situation confuse. Nous proposons une équation LFER et son échelle de similarité quadriparamétrique (ICAD), obtenue par analyse statistique. Pour chaque Xi, un paramètre rend compte des effets inductifs (I) et trois autres de ceux de délocalisation des électrons par résonance. La modélisation avec cette échelle (ICAD) nous a permis de mieux rendre compte de nombreuses situations de la spectrométrie et de la réactivité moléculaire que les LFER complexes (YT, CHARTON, TAFT).

1 Local and/or Unique Scales: Simple and Complex LFER

Hammett's LFER relation laid the foundation for a quantitative scale of group or atom effects considered as active substituents (X_k) of some XGY structures. This equation was excellent for expressing certain similarities between two kinds of comparable reactions or properties. It enables the establishment of σ_j (ScX_k) parameter scales, measuring the action of X_k groups used in simple linear free energy relationships (LFER).

1.1 Local and/or unique Scales

In practice the success of this proposition could have led to one or more general tools capable of extracting "numerous limited scales" but precisely adapted to certain properties, for a YG_k support structure. Thus, chemical complexity would be approached by a plurality of similarity expressions. In a different approach, some researchers wanted to obtain a "unique scale", i.e. a very general one and, as a corollary, looked for certain kinds of new general similarity equations. This second and very different road led to the fusing together of excellent but limited results in order to obtain "unique scales" unquestionably broader in scope but often far less precise.

1.2 Complex similarity LFER

Frequently, the statistical correlation search was broadened by proposing "more interpretative" equations oriented so as to use the physical factors on which experimental similarities were supposedly based. Many attempts were thus made to generate structure property equations with two apparent parameters, with limited applications. De facto, this approach gradually justified the use of similarity equations with several independent factors, some of which are linked and/or limited to well defined fields of applications [3-6].

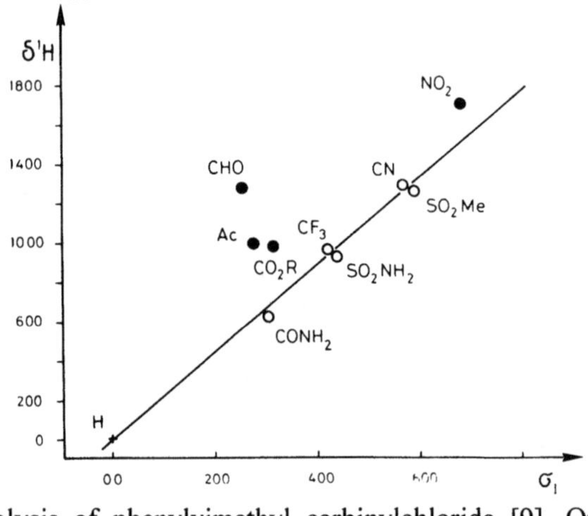

Figure 1. Solvolysis of phenylvimethyl carbinylchloride [9]. Only the MeO-like groups (•) deviate. The halogens show only the enhanced polar resonance. Thus there are two different classes of releasing substituents, leading to the definition of a new donating scale (σ_d) besides the more usual (σ_c) scale.

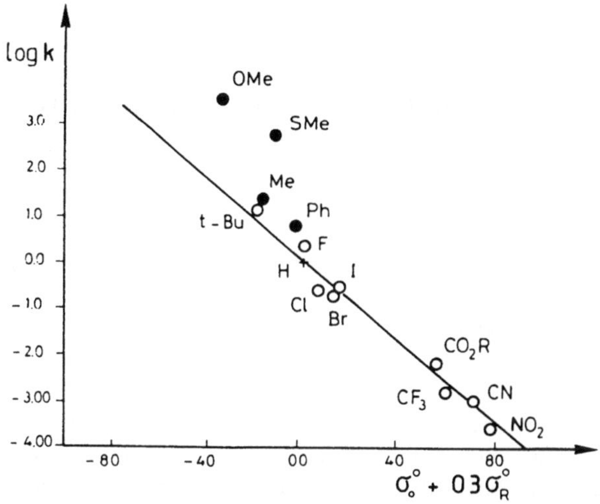

Figure 2. NMR of the OH proton in p-XC6H4OH [10]. The RCO-like groups (•) deviate. Thus there are two classes of withdrawing substituents, involving the need for a specific accepting scale (σ_a).

For our part we have tried to consider the diversity of physico-chemical situations in proposing a new breakdown of basic contributions of X_k group effects. We have identified the possible combinatory action of four interaction mechanisms for interpreting the global effects of X_k groups. Two of these effects are well known (σI and σR°). The need for two more effects (corresponding to through-resonance) is demonstrated in Figure1 and Figure 2.

1.3 Normative conditions

Practically speaking, instead of trying to manage complexity by merging earlier data, we handled many results known to date, integrating into our statistical analysis, not only our hypotheses on the four interaction parameters, but also those elements of classical standardization required for a very broad statistical analysis. For this study, carried out on a large number of data, we freed ourselves from the many, often implicit, hypotheses proposed in turn during the saga of the ScX_k [1][7]. We have indeed spelled out new and very precise normative conditions that justify our comparison of the many scales in the literature.

In this article we do not attempt a critical study, carried out elsewhere, of similarity as seen by Hammett and his followers. We avoid evaluating Taft type biparametrical equations and rather discuss those pertinent and usual equations that call for complex

[1]The ρ_m=ρ_p relation, for example, is one of these uncontrolled hypotheses.

handling of parallel mechanisms of interaction (YT and Charton). We present the results obtained from our original ICAD equation, giving a partial view of our scale (σ_i, σ_c, σ_d, σ_a). Examples of applying our ICAD method to six areas of exploitation are presented and compared with the results of regressions belonging to other models (Taft, YT and Charton).

2 The Modeling of Substituent Effects: A New Approach

As we have seen, the model will be:

$$Q_X - Q_H = \rho_i\ \sigma_i + \rho_c\ \sigma_c + \rho_a\ \sigma_a + \rho_d\ \sigma_d$$

This is a four factor model with four adjustable parameters. It represents clearly the known classes of reactions. However, in the chemical reaction or equilibrium area, at least one of the parameters will generally be null, depending on the Y reacting center and the G transmitting system. More complex models are found only in such areas as NMR chemical shifts or in complex chemical kinetics. In the field of LFER, usually formulated scales are based on arbitrary primary choices of some ρ_j and the primary or secondary scales are improved by successive iterations of σ_j and ρ_j factors or parameters in turn. The situation is somewhat confused with regard to complex and parallel development. In elaborating our four parametric scale, we have proceeded by standard procedures whose essential aspects are given below.

2.1 Scale definition; modeling rules and σ_j signs (σ_j arbitrary polarities)

The basic scales are usually defined through the adoption of a reference compound (X_k = H) and a set of reference reactions (for instance, ρ = 1 for the pK of benzoic acids in H_2O at 25°C). These definitions lead to primary and secondary scales with many unsolved or arbitrarily solved problems.

For our approach, we chose two reference substituents for each scale. This choice allows the use of all available series for the definition of each substituent constant. In our choices, we defined values compatible with the most popular substituent scales. Our scaling references are the following:

$$X_k = H\ (\sigma_i = \sigma_c = \sigma_d = \sigma_a = 0),$$

$$CN\ (\sigma_i = 0{,}57),\ F(\sigma_c = 0{,}437),$$

$$OMe\ (\sigma_d = 0{,}423),\ Ac(\sigma_a = 0{,}149).$$

It is also important to consider the arbitrary signs of σ_j parameters and ρ_j selectivity constants. These are mutually dependent. Here we chose to give positive values to the resonance group factors. This implies that the corresponding ρ_i, ρ_a, ρ_d reflect this decision. In most cases, ρ_i and ρ_a (which are measures of electron withdrawing effects) share the same sign, while ρ_c and ρ_d (electron releasing effects) share the opposite sign. Differences in the ρ_j values will be revealed when comparing with other models.

2.2 Iterative elaboration of our ICAD scale

The four subsets of the substituent factors are obtained by an iterative process, considering the profiles of the available sets or series.

Table I: The ICAD Scale

	Inductive σ_i	Conjugative σ_c	Accepting σ_a	Donnating σ_r
INDUCTIVE SUBSTITUENTS				
H	0	0	0	0
CN	0.57	0	0*	0
SO_2Me	0.596	0	0	0
SO_2NH_2	0.52	0	0	0
SO_2Ph	0.552	0	0*	0
CF_3	0.423	0	0	0
ACCEPTING SUBSTITUENTS				
NO_2	0.68	0	0.086	0
$CONH_2$	0.273	0	0.094	0
COOH	0.307	0	0.129	0
COOEt	0.313	0	0.135	0
COMe	0.32	0	0.149	0
CHO	0.381	0	0.241	0
CONJUGATIVE SUBSTITUENTS				
SCN	0.549	0.163	0	0
OCF_3	0.503	0.194	0	0
Br	0.46	0.264	0	0
Cl	0.46	0.282	0	0
F	0.503	0.437	0	0
DONATING SUBSTITUENTS				
I	0.426	0.22	0	0.028
Me	-0.005	0.065	0	0.146
t-Bu	-0.012	0.056	0	0.16
Ph	0.125	0	0	0.195
OAc	0.366	0.14	0	0.24
OMe	0.27	0.37	0	0.423
OPh	0.363	0.315	0	0.424
SMe	0.283	0.208	0	0.431
OH	0.243	0.347	0	0.478
NHAc	0.255	0.229	0	0.505
NH_2	0.15	0.354	0	0.872
NMe_2	0.159	0.353	0	1.195

The inductive scale is derived first, using the relevant available series. We start with a usual σ_i scale (TAFT). The conjugative resonance set (σ_c) is obtained by the series with a weak direct resonance effect. The donating and accepting scales (σ_d and σ_a) are derived, using the corresponding reaction series (the starting scales were derived graphically from σ_p^+ and from the acidity of para phenols) [11]. Indeed the physical significance of σ_i and σ_c constants are not different from that of σ_I and $\sigma_R°$. The physical meaning of σ_d resembles that of $\Delta\sigma^+$. On the contrary, the physical meaning

*slightly positive value is possible

of σ_a is very different from that of σ^- [11]. σ_d and σ_a reflect the two types of through-resonance donor D towards acceptor A : D → A and A ← D in XGY.

Some substituent sets are given in Table I. They show the four substituent types thus obtained, easily recognized by the sets of zero values. This allows for a good definition of experimental designs to study the substituent effects. The interpretations of the ρ_j values is facilitated and the conditions for good modeling are easily found.

TABLE II
Efficiency of MODELING

TAFT	**TA**	*	ρ_F	ρ_R	ρ_R^+	$\rho_Ø$	*					
YUKAWA-TSUNO	**YT**	*	ρ_I	$\rho_{R°}$	$\rho_°^+$	$\rho_°^-$	*					
CHARTON	**MC**	*	ρ_L	ρ_D	ρ_E	–	*	r^2	s_r	**F**	**n**	
THIROT	**TH**	*	ρ_i	ρ_c	ρ_r	ρ_a	*					
	TA	*	-4.8	-29.7		2.38	*	0.97	1.5	202	23	
p-XC_6H_4F	YT	*	-5.72	-19.8	-5.63	-6.19	*	0.98	1.2	219	26	
	MC	*	-7.65	-23.6	0		*	0.97	1.3	***389***	27	**5d**
RMN ^{19}F	TH	*	-13.9	33.7	5	-14.7	*	0.98	***1.1***	331	28	
	TA	*	4.92	9.49		-6.95	*	0.99	0.97	113	8	
p-XOC_6H_5	YT	*	16.56	0	-6.85	0	*	0.94	1.66	58	9	
	MC	*	11.88	0	-30.2		*	0.99	0.8	260	9	13
RMN $^{13}C_p$	TH	*	14.62	0	12.28	15.87	*	0.998	***0.34***	***951***		
	TA	*	0	12.03		-1.58	*	0.99	0.45	389	14	
p-$XC_6H_4NMe_2$	YT	*	0	2.85	5.4	5.93	*	0.99	0.38	359	14	
	MC	*	2.14	9.56	0		*	0.99	0.4	503	14	8
RMN $^{13}C_p$	TH	*	3.61	-11.6	-2.65	1	*	0.998	***0.21***	***907***	14	
	TA	*	-3.06	-3.54	0	0	*	0.94	0.37	67	11	
p-XC_6H_4OH	YT	*	-3.23	0	-26.91	0	*	0.82	0.64	20	11	9
	MC	*	-3.15	-2.69	0		*	0.94	0.37	67	11	
pK(H_2O)	TH	*	-3.57	3.55	0	-2.85	*	0.97	***0.26***	***94***	11	
	TA	*	-6.3		-8.34	0	*	0.99	0.28	343	9	
p-$XC_5H_5NH^+$	YT	*	-7.35	-8.05	5.52	0	*	0.992	0.28	236	9	
	MC	*	-6.41	-4.62	-13.32		*	0.995	0.2	448	9	**2b**
pK(H_2O)	TH	*	-6.33	4.98	3.39	2.4	*	0.998	***0.15***	***603***	9	

2.3 Applications

Some results are given to show the enhancements thus obtained, when compared with the YT or MC modeling (Table II).

The series given in Table II were chosen (from more than 2000) to cover the full range of properties (physical [NMR ^{19}F and ^{13}C], reactivities, structures).

Far better results were obtained with our modeling, as is shown by the corresponding standard error for regressions and Fisher's values.

CONCLUSION

We were able to obtain a good four parametric scale for describing electrical effects of substituents because:

- We used a classification of the substituents that accurately reproduces their qualitative behaviour.
- We coherently and systematically excluded the erroneous hypotheses and the poor definition of constants from the model.
- σ_i measures the inductive effect. This contribution is always more or less present.
- The ρ_i^p / ρ_i^m ratio gives clues to the resonance of the reaction center in aromatic structures. This is easily extended to all aromatic or heteroaromatic structures.

The method provides reaction constants with easily understood meanings:

- ρ_c measures the conjugation between an unsaturated G structure and a substituent attached to it (polar resonance effect).
- ρ_d measures the direct resonance of a donating substituent and an electrophilic center.
- ρ_a measures the direct resonance of a withdrawing substituent and a nucleophilic center.

The above shows clearly that we cannot obtain a complete modeling if in a reaction series, one scale of substituents is not represented. Unfortunately this occurs very often in the literature.

Does modeling aim only at the prediction of values not experimentally determined? We do not believe so. In one of his papers [12], Exner wrote "without the interpretation of the results in chemical terms, the correlation analysis would become a play with numbers". Though interpretations and correlations of data can and are often separately successful, we fully agree with this statement which indicates that the main role of modeling is to increase the understanding of chemical and physical phenomena.

KEYWORDS

Substituent scale, Electrical effects, Yukawa-Tsuno, Charton, ICAD model, Complex LFER.

REFERENCES

[1] Yukawa Y., Bull.Chem.Soc.Jap., 93, 4187 (1971)
[2] Charton M.,
a. Proc.Chemometrics, 20 (1990)
b. Prog.Phys.Org.Chem., 16, 287 (1987)
c. Stud.Org.Chem., 31, 537 (1986)
[3] Nieuwdorp G.H.E., J.Chem.Soc., PT2, 537 (1979)
[4] Happer D.A.R., J.Chem.Soc., PT2, 694 (1979)
[5] Taft R.W.,
a. J.Am.Chem.Soc., 74, 3120 (1952)
b. J.Phys.Chem., 64, 1805 (1960)
c. Prog.Phys.Org.Chem., 16, 1 (1987)
d. J.Am.Chem.Soc., 85, 3146 (1963)
[6] Bromilow J., J.Org.Chem., 45, 2429 (1980)
[7] Hine J., J.Am.Chem.Soc., 81, 1126 (1959)
[8] Bijloo G.J., Quant.Act.Rel., 2, 124 (1983)
[9] Ehrenson S., Prog.Phys.Org.Chem., 10, 1 (1973)
[10] Tsuno Y., Bull.Chem.Soc.Jap., 45, 1519 (1972)
[11] Thirot G., Bull.Soc.Chim.Fr., 739 (1967)
[12] Exner O., Org.React. (TARTU), 16, 285 (1979)
[13] Bradamante S., J.Org.Chem., 45, 114 (1980)

THE OASIS COMPUTER SYSTEM AND ITS NEW TOOLS FOR QSAR STUDIES

***Ovanes G. MEKENYAN**[a,b,1] **Julian M. IVANOV**[b], **Stoyan H. KARABUNARLIEV**[b] **and Dimitar N. DIMITROV**[b]*

[a]*Lake Superior Research Institute, University of Wisconsin, Superior, WI 54880 USA*
[b]*Higher Institute of Chemical Technology, 8010 Bourgas, Bulgaria*

ABSTRACT

The revised OASIS system for computer assisted quantitative structure-property analysis is much more flexible and versatile than the previously introduced version. The most significant contribution is the newly developed QSAR methodology, named "dynamic" for selecting the active conformers and significant stereoelectronic parameters. It is assumed that in general the 3D molecular shapes result from an integral effect of multistage interactions carried out in various solvent environments in which different conformers could be active. This conformer multiplicity is not accounted for by the conventional QSAR methods. Our new methodology core is based on an algorithm for exhaustive molecular design and a system for interactive conformer screening, whose rules are based on chemical expertise, stereoelectronic parameters and interaction mechanisms. Dynamic QSAR often allows one to omit the time consuming conformer geometry optimization procedures.

RESUME

Le système OASIS , nouvelle version , pour l'analyse de structure propriété par assistance informatique est plus souple et versatile que l'ancienne version. La contribution la plus importante consiste en une méthodologie QSAR, dite "dynamique" pour la sélection des conformères actives ainsi que ceux des paramètres stéréoélectroniques significatifs. Cette méthode dite du type "mimique" rend compte de la multiplicité des formes moléculaires 3D des composés chimiques. Elle repose sur l'hypothèse d'une intégration des effets des interactions. La multiplicité des conformères n'est pas prise en compte par les méthodes conventionnelles QSAR , où la structure 3D porte vers le comformère d'énergie plus basse. La méthodologie proposée est exhaustive mais suivi d'un système interactif de criblage. Le "QSAR dynamique" proposé permet dans de nombreux cas de raccourcir beaucoup les durées

---

[1]1 To whom correspondence may be addressed

Running Head: OASIS system and dynamic QSAR method

usuelles d'optimisation des calculs de géométrie conformationnelle. Souvent on peut éviter ce type de calcul.

1 INTRODUCTION

The existing (conventional) QSAR computer systems provide 3D structure builders proceeding from molecular topology. Details for most applied 2D-3D builders (converters) can be found in ref. 4 and 5. The common feature of all 2D-3D model builders is their ability to arrive eventually at a single good quality stereoisomer corresponding to molecular topology input. In addition, techniques [4, 6] are used to explore the conformational space to further improve quality of the 3D model (i.e. reaching the minimum energy conformer). Usually such methods are combined by force field packages for fast assessment generated by conformers' energy. Exploring the conformational space, these techniques arrive either at the lowest energy conformer or at a set of the most stable conformers. Eventually, within the available conventional QSAR methods, one selects a molecular structure that represents the most stable (single) conformer whose geometry is additionally optimized by quantum-mechanical (or force field) method. However, the most stable conformations of a molecule are not always the active ones. Thus for example, because of the requirement for an antiparallel arrangement of the leaving groups, the less stable axial conformer of cyclohexyl bromide (or p-toluenesulphonate) undergoes the elimination (E2) reaction, instead of the more stable equatorial conformer.

Analogous principles hold for enzymatic reactions, where enzymes may change their conformations either by activation through external molecules or by contact with the substrate, according to the "allosteric" effect. Moreover, the available force field or quantum-chemical calculations are applicable under the basic approximation of the "isolated" molecule. Thus, they do not take into account the neighbouring effects of other molecules including those of solvents.

In contrast, the recently introduced dynamic QSAR method [8] attempts to take into account the multiplicity of conformers taking part in interactions carried out in complex reaction environments, depending on the particular interaction step. The new method is implemented in the new version of OASIS computer system [3]. It is based on two general algorithms. The 3DGEN system [4], which proceeds from molecular topology and generates all conformers fulfilling some requirement for steric constraints and chemical expertise, requires the user to formulate general hypotheses for the structure of the active conformers. The structural constraints inherent in the 3DGEN system include parameters such as the distances between non-bonded atoms, ring-closure parameters and torsional resolution which, when specified by the user, prevent the generation of unlikely high energy conformers. In addition, one can impose a chemical expertise requiring the design of specific conformers. For example, those conformers which can assume a near planar structure, or those with hydrogen bonding alignments, or specific enantiomers can be identified.

Thus, one arrives at the first broad set of conformers corresponding to some initial user's hypotheses. Next, the stereoelectronic structure of each of the conformers is assessed by quantum chemical methods. Usually MOPAC6 package [9] is applied in combination with the OASIS routine for stereoelectronic indices calculation. The molecular structure descriptors can be determined either after geometry optimization or by single point (1SCF) calculations. However, we have often found greater variation in indices among different conformers of a given molecule than between some conformers of different molecules.

In many instances, at least when highly strained structures are not involved, the stereoelectronic indices obtained without geometry optimization may be sufficient for estimating chemical reactivity within the dynamic QSAR method.

The conformer screening routine is the other essential algorithm of the dynamic method. Conformer distributions based on energy or any other selected geometric, physico-chemical, and/or stereoelectronic parameter can be examined and then used in the search for a reasonable, mechanistically relevant QSAR. There are a variety of schemes designed to examine the distribution of values among the conformers. For example, the distribution of each parameter can be displayed among a selected number of parameter windows, each having an equal percentage of the whole parameter range. The conformers belonging to the most populated window are considered to provide a **prevailing range of parameter values** (Fig. 1).

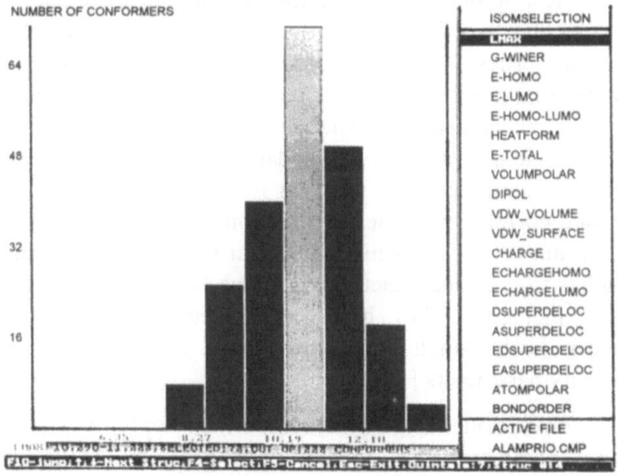

Figure 1. Selected conformers for 2-decyn-1-ol (72 out of 222) with prevailing values of the maximum distance L_{max} in the range 10.290 Å to 11.223 Å. The introduced number of quintals is seven (see ref. 8 for more detail).

Similarly, the number of conformers examined can be diminished by selecting a single conformer from each window, thus providing reduced representation of the **whole range of parameter variation**. Next, conformers can be selected having

extreme (maximum or minimum) values of the parameter and which populate the outermost windows of the parameter distribution. Alternatively, relying on calculated heat of formation of each conformer and using Boltzman's distribution formula, one can select conformers that **populate a certain percentage of energy levels.**

A **parameter weighting** algorithm is also available according to which the parameters for each chemical can be normalized with respect to the statistical weights of the constituting conformers.

The above conformer generation and selection system enables specific hypotheses to be incorporated for the mechanism of interaction and tests them with a more realistic subset of 3D conformers. The latter often do not correspond to the low energy states of the chemicals. In the next QSAR analysis, the potentially active conformers are considered as distinct chemicals, and characterized by the value of the dependent parameter of the parent (2D) compound but with different stereoelectronic parameters determined by their 3D molecular shape.

2 SELECTION OF THE SIGNIFICANT PARAMETERS WITHIN THE OASIS SYSTEM

Large numbers of descriptors can be incorporated in the parameter pool after the assessment of molecular two dimensional and stereoelectronic structures as well as physico-chemical properties. The review of QSAR studies, published in the last few years, clearly shows that proliferation of molecular indices has expanded well beyond our ability to ascribe mechanistic relevance to the indices [10]. The tendency toward proliferation of molecular descriptors is a dangerous bias which may also cause difficulties due to the combinatorial problem in statistical analysis and the risk of arriving at a chance correlation. The construction of the descriptor pool should be based on a limited number of mechanistically clear concepts. For example, there are different approaches to chemical reactivity, such as the static [11], localization [12], delocalization [13] methods. They, however, are generalized by the polyelectronic perturbation theory [14], which alone provides a reliable theoretical basis for introducing reactivity parameters [10].

In contrast to the conventional QSAR methods, which apply the statistical approach for the screening of parameters from the descriptor pool, the OASIS system employs the statistical tools after the reduction of the initial set of molecular descriptors by non-formal (physico-chemical) principles. The latter implies involvement of the users' hypotheses concerning interaction mechanism(s). For example, if one anticipates that charge transfer processes control the interaction under consideration, then the set of molecular descriptors can be strongly reduced to a narrow set of stereoelectronic descriptors, such as the energies of frontier orbitals and their difference, as well as certain local reactivity parameters, such as superdelocalizabilities, frontier charges, etc. Accordingly, many structural (i.e.

topological and volumetric indices) and physico-chemical parameters (logP, molar refraction, boiling point, etc.) should be eliminated from the data matrix, thus reducing in advance its dimensionality.

After the selection of potentially significant descriptors, one should employ a statistical analysis in order to derive the ultimate QSAR equation. Our experience shows that if the rigorous statistical criteria are applied during model derivation the incorporation of more than two to three parameters in the ultimate correlation is not likely. Thus, penetration factor (as described by logP) and soft electrophilicity (assessed by superdelocalizabilities and frontier orbitals) was found to explain the acute toxicity variation of industrial chemicals [15, 17.] AChE inhibition of organophospates (OPs) was also modeled [10,18] by hydrophobicity and hard reactivity parameters (as estimated by the charge at P-atom or bond order of anhydride ester bond). Orthogonality found between reactivity and hydrophobicity parameters supported the concept of the "3D space" toxicity model [10, 15, 17]. According to this general QSAR, in modeling compounds with large structural variations, high correlations between toxicity and each of the individual factors which condition it should not be expected. However, it is expected that toxicity will be described by a high quality plane or other surface illustrating the complementary or combined effect of the significant factors. On the other hand, orthogonality between factors should also be considered as competition between them. Thus, some highly reactive chemicals could be of low toxicity if the structural fragments providing high reactivity (such as -F, $-NO_2$) decrease their penetrability at the same time [10].

At the same time, valid QSARs may be overlooked if the focus is on linear relationships only. Especially when there are competing interaction steps, multi-phase relationships with pronounced extrema should be expected. For example, one and the same reactivity parameters were found [10] to condition both the persistency (alkaline hydrolysis) of OPs and their toxicity (AChE-inhibition) which eventually resulted in a parabolic relationship with extremum. In a series of papers [19, 22] it was shown that the photoinduced toxicity of polycyclic aromatic hydrocarbons (PAHs) is a result of competing effects including structural (stability and light absorbance as assessed by HOMO-LUMO gap) and irradiation (light energy and intensity) factors.

3 APPLICATIONS OF THE DYNAMIC QSAR METHOD

The efficacy of the dynamic approach for active conformer selection is illustrated by the following examples:

3.1 Semicarbazide-Induced Mortality in Frog Embryo

The acute lethality of 36 semicarbazides and thiosemicarbazades, evaluated using the Frog Embryo Teratogenesis Assay: *Xenopus*, were modeled by using the dynamic

QSAR method [23]. The assumed mode of action, osteolathyrism, was defined by the failure of connective tissue to polymerize properly due to interference with lysyl oxidase. Active conformer selection was based on 3D isomer distribution according to: the frontier orbital energies and volume polarizability, conditioning their reactivity and hydrophobicity, respectively. The best results were obtained by the selection of conformers (n=110) that provided prevailing values of electron acceptor properties. Moreover, the two best QSAR parameters encompassing all the evaluated compounds, incorporate a steric parameter, the geometric analog of Wiener topological index, and the local electronic characteristics of the C=O or C=S group, superdelocalizabilities and charges. The best QSAR model (Fig. 2) for FETAX acute lethality has the following statistics: n = 110; r^2 = 0.805; s^2 = 0.21; F = 220.8; = 99; r^2_{cros} = 0.812; s^2_{cros} = 0.27. The conventional statistics were extended by the cross-validation test estimates: cross-validated r^2 and s^2. It was found that toxicity increased with the increase of the size of substituents (probably also related to the penetration factor), increase in the electron acceptor properties of carbonyl oxygen or sulfur, and the positive charge at the carbonyl carbon. Moreover, acute lethality was found to be conditioned by a single action mechanism. The results were obtained without geometry optimization of the generated conformers.

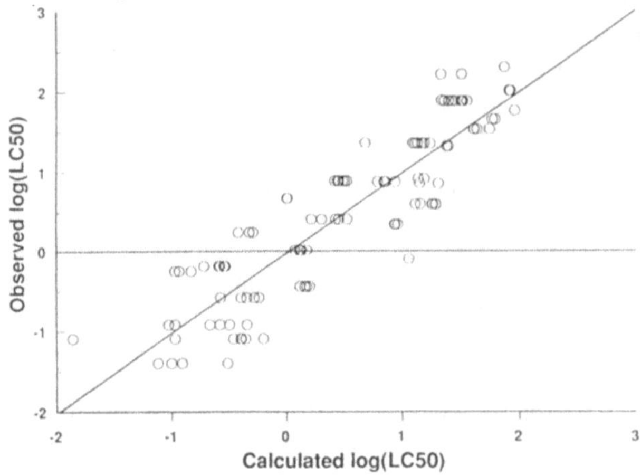

Figure 2. Plot of observed ($\log(LC_{50})$) versus calculated embryo lethality data.

3.2 AhR Binding Affinity of PCBs

Different schemes for deriving QSAR for aryl hydrocarbon receptor (AhR) binding (log(1/EC50) of 14 polychlorinated biphenyls (PCBs) were studied [24-25] (group A). The 3DGEN system generated 103 conformers under the steric constraints of 1.75 Å non-bonded cutoff (Van der Waals radii of Cl) and 60° torsional resolution. A stacking type of interaction with AhR was hypothesized and the receptor affinity of PCBs was correlated with the energies of frontier (LUMO and HOMO) orbitals. The next QSAR study was based on non-optimized conformers whose electronic structure was assessed by single point (1SCF) calculations. As can be seen from Figure 3a, the lack of optimization resulted in a large variation of E_{LUMO} for the conformers of each chemical and ultimately no correlation was found between log(1/EC50) and the electronic parameter ($r^2 = 0.02$). During the 1SCF calculations conformer geometries were assumed as fixed. In order to prevent the presence of highly unstable conformers it is recommended to keep in the correlation sample only those conformers populating 80% of E-levels (Fig. 3b). The elimination of high energy conformers did not result in a statistically satisfactory equation, $r^2 = 0.26$. The selection of most planar conformers, however, yielded a significant QSAR model, $r^2 = 0.78$ (Fig. 3c). Similar results were obtained after geometry optimization of conformers. The correlation with the energetically most favorable conformers was low, $r^2 = 0.43$. However, the selection of most planar conformers significantly increased the variance of the model: $r^2 = 0.72$.

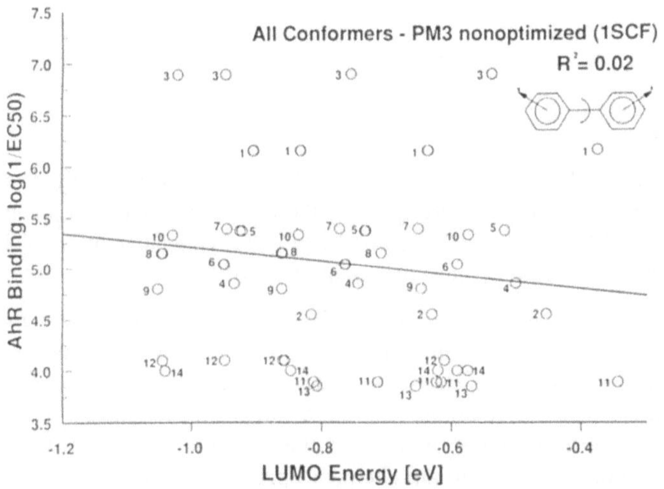

Figure 3a.

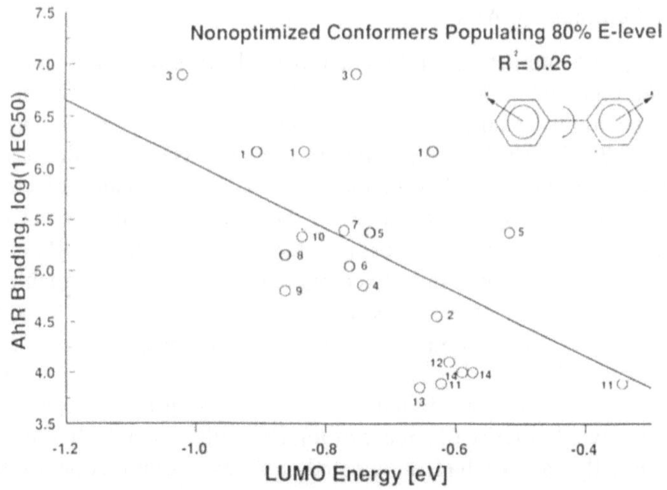

Figure 3b.

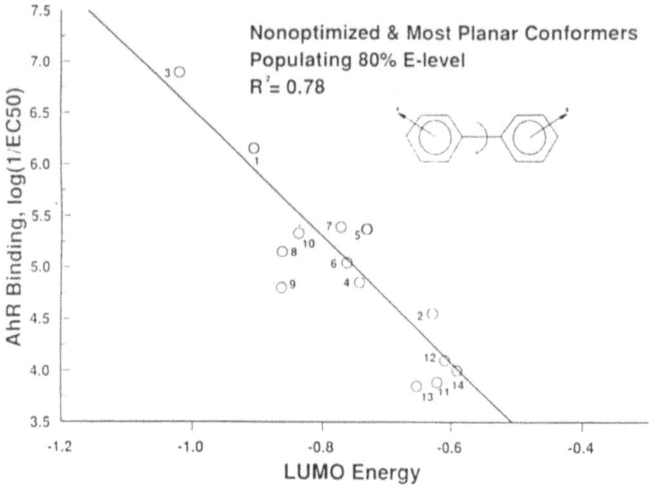

Figure 3c.

Figures 3a, 3b, 3c. Plot of observed (log(1/EC50) AhR binding data versus E_{LUMO} for non-optimized PCBs of group A. In the correlation samples are included: (a) all conformers; (b) the most stable conformers occupying 80% of E-levels; (c) the most planar conformers among those occupying 80% of E-levels.

Apparently, the active conformers of PCBs of group A taking part in the AhR receptor binding are the most planar rather the most stable ones, which supports the hypothesis for the stacking type of interactions. After the elimination of the extremely unstable conformers, 1SCF calculations are quite sufficient to assess the stereoelectronic structure of conformers with respect to the receptor binding. The time consuming geometry optimization (quantum-chemical) procedures did not improve the quality of the derived ultimate QSARs.

The cleft type of interactions and related local electron-acceptor (donor) parameter of the second biphenyl moiety (substituted by fragments of different polarity) were found to be highly significant for a second class of PCBs incorporating 16 4'-substituted-2,3,4,5-tetrachlorobiphenyls (group B). Based on the assumption that a charge-transfer is performed from the amino acid-side chains of the AhR towards PCBs, we found higher (lower) values of the local electron-acceptor (donor) indices or positive charges at the lateral positions for the more active PCBs. The compounds from this class have a structurally fixed biphenyl ring (2,3,4,5-tetrachloro) providing a "constant" contribution of the stacking type of interaction, which explained the low statistical weight for E_{LUMO}, E_{HOMO} and $E_{HOMO\text{-}LUMO}$ parameters that directly describe these kinds of interactions. In accordance with the large variation of the substituent polarity, we found logP to be a significant regressor, so that the increase of solvent polarity (decrease of log P) resulted in a decline in binding affinity (Fig. 4). Finally, as we had anticipated, a linear combination of the parameters controlling the stacking and cleft type of interactions explained the variance of toxicity for the combined classes of PCBs (Fig. 5).

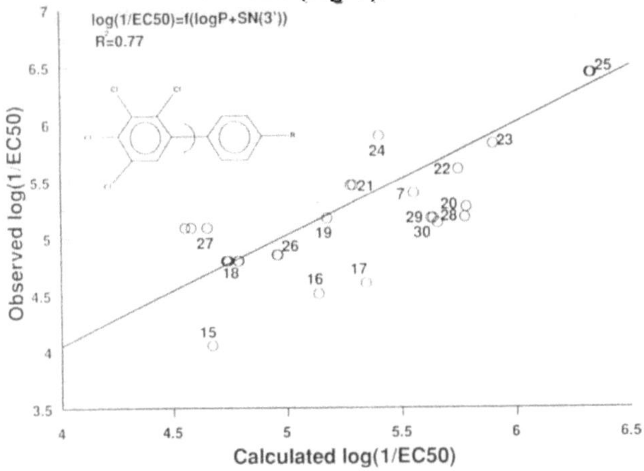

Figure 4. Plot of observed (log(1/EC50) versus calculated AhR binding data for non-optimized PCBs of group B.

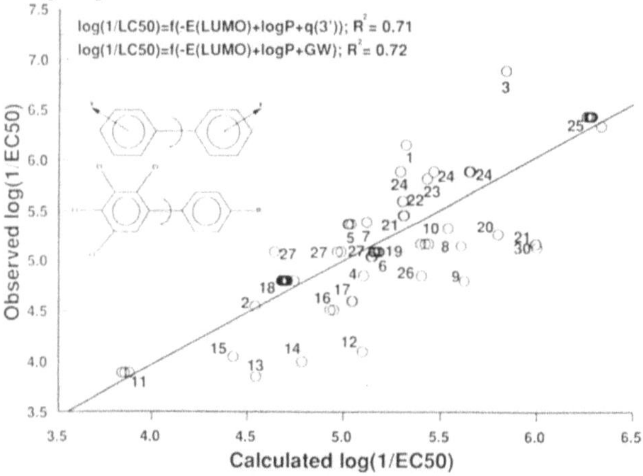

Figure 5. Plot of observed (log(1/EC50) versus calculated AhR binding data for optimized PCBs of combined group AB.

3.3 Photoinduced Toxicity of Thiophenes

The generality of the multilinear (parabolic) relationship between phototoxicity and the electronic gap established for rigid PAHs was examined for flexible aromatic chemicals such as naturally occurring and synthetic tricyclic thiophenes [22]. An initial set of 812 conformers was generated to represent the 41 chemicals from the thiophene series. It was found that the ranges of calculated HOMO-LUMO energy gaps for different conformers of a single chemical are significantly large (in some cases larger than 1 eV). This unambiguously showed the necessity of applying the dynamic QSAR approach. The screening of the initial set of 812 conformers was performed according to: (i) heats of formation (formation enthalpies); (ii) volume polarizability, VolP, which is directly related to the octanol/water partition coefficient; and (iii) HOMO-LUMO gap, the ground state electronic parameter which dictates the light absorbance and stability of PAHs relative to their phototoxicity. The conformers found to develop parabolic phototoxicity/electronic gap models were the energetically most favorable ones, for both optimized and non-optimized isomers (Fig. 6).

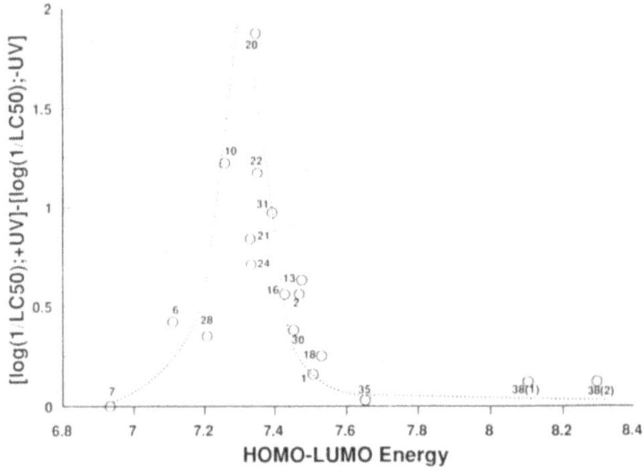

Figure 6. Variation of photoenhanced toxicity (log(1/LC50)) for mosquito larvae with HOMO-LUMO gap of thiophenes, for geometry optimized conformers occupying 10% of the energy levels.

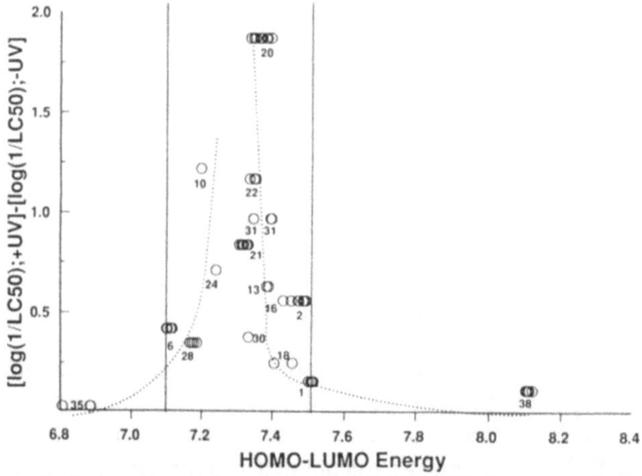

Figure 7. Variation of photoenhanced toxicity (log(1/LC50)) for mosquito larvae with HOMO-LUMO gap of thiophenes, for geometry optimized conformers with minimum HOMO-LUMO gap.

Other conformers creating a biphasic relationship are those with minimum electronic gap among optimized isomers (because of their high "red-shiftiness", Fig. 7) and conformers providing the prevailing range of the gap among non-optimized isomers (due to their high statistical weight with respect to the HOMO-LUMO gap).

The range of the phototoxic region along a HOMO-LUMO energy gap axis for optimized thiophene derivatives (within the PM3 hamiltonian) coincides almost exactly with that established for PAHs. Its left side boundary is around 7.1 eV, whereas the right side is around 7.5 eV. For non-optimized conformers this region is shifted toward higher values of the electronic gap due to structural effects not accounted for in 1SCF calculations.

CONCLUSION

The dynamic QSAR method implemented in the OASIS computer system deals with the fact that in complex environments such as biological tissues and fluids, the molecules are likely to exist in conformations other than the lowest gas phase energy state. Hence, the chemical behavior (activity, toxicity, reactivity, etc.) may be an integral effect of a set of conformers rather than a property of a single, low energy conformer. In this respect, searching for molecules that can assume a specific conformation may be as important as searching low energy conformers for possible activity. If poor correlations were found between chemical activity and a mechanistically plausible stereoelectronic index, it seems as important to consider models with conformations other than those of lowest energy as it would be to search for different stereoelectronic indices. The results show that the search for likely active conformers is in general more important than determination geometry of the most stable 3D isomers. This is reasonable after finding that the variation in indices among different conformers of a given molecule is often larger than that between conformers of different molecules. Accordingly, the indices of electronic structures, without geometry optimization may in many case be sufficient for estimating chemical activity. If so, delaying geometry optimization until necessary would save a significant amount of computing time and facilitate the modeling process.

KEYWORDS

QSAR, Conformer generation, Conformer activity, Interactions.

REFERENCES

[1] Mekenyan O., Karabunarliev S., Bonchev D., The OASIS Concept for Predicting the Biological Activity of Chemical Compounds, J. Math. Chem., 4, 207-215 (1990)

[2] Mekenyan O., Karabunarliev S., Bonchev D., The Microcomputer OASIS System for Predicting the Biological Activity of Chemical Compounds, Comput. Chem., 14, 193-200 (1990)

[3] Mekenyan O.G., Karabunarliev S.H., Ivanov J.M., Dimitrov D.N., A New Development of the OASIS Computer System for Modeling Molecular Properties, Comput. Chem., 18, 173-187 (1994)

[4] Ivanov J.M., Karabunarliev S.H., Mekenyan O.G., 3DGEN: An algorithm for exhaustive generation of 3D isomers proceeding from molecular topology, J. Chem. Inf. Comput. Sci. 34, 234-243 (1994)

[5] Gordeeva E.V., Katritzky A., Shcherbukhin V.V., Zefirov N.S. Rapid Conversion of Molecular Graphs to Three-Dimensional Representation Using the MOLGEO Program, J.Chem. Inf. Comput. Sci., 33, 102-111 (1993)

[6] Lipton M., Still W.C., J. Comput. Chem., 4, 343 (1988)

[7] Eliel E.L., Chemistry in Three Dimensions, in Chemical Structures, W.A. Warr (Ed.) Springer, Berlin (1993)

[8] Mekenyan O.G., Ivanov J.M, Veith G.D., Bradbury S.P. DYNAMIC QSAR: A methodology for investigating structure-activity relationships, Quant. Str.-Act. Relat., 13, 302-307 (1994)

[9] Stewart J.J.P. MOPAC6, Frank J. Seiler Research Laboratory, U.S. Air Force Academy, CO 80840 (1990)

[10] Mekenyan O.G., Veith G.D., The Electronic Factor in QSAR: MO-Parameters, Competing Interactions, Reactivity and Toxicity, SAR and QSAR Environ. Res. 2, 129-143 (1994)

[11] Higasi K., Baba H., Rembaum A., Quantum Organic Chemistry, John Wiley & Sons, New York, pp. 262-340 (1965)

[12] Streitwieser, A., Molecular Orbital Theory for Organic Chemists, John Wiley & Sons, New York, p.258 (1961)

[13] Klopman G., Chemical Reactivity and Reaction Paths: General Introduction, in Chemical Reactivity, G. Klopman (Ed.) John Wiley & Sons, New York, p. 1-12 (1974)

[14] Klopman G., The Generalized Perturbation Theory of Chemical Reactivity and its Applications, in, Chemical Reactivity and Reaction Paths, G. Klopman (Ed.) John Wiley & Sons, New York, p. 55-166 (1974)

[15] Mekenyan O., Veith G., Bradbury S., Russom C., A QSAR Approach to Estimating the Toxicity of Unsaturated Alcohols to the Fathead Minnow, (Pimephales Promelas), Quant. Str.-Act. Relat. 12, 132-136 (1993)

[16] Mekenyan O.G., Veith G.D., Relationships Between Descriptors For Hydrophobicity and Soft Electrophilicity in Predicting Toxicity, SAR and QSAR Environ. Res. 1, 335-344 (1993)

[17] Veith G.D., Mekenyan O.G., A QSAR Approach for Estimating the Aquatic Toxicity of Soft Electrophiles [QSAR for Soft Electrophiles], Quant. Struct.-Act. Relat. 12, 349-356 (1993)

[18] Schurmann G., QSAR analysis of the acute fish toxicity of organic phosphorothionates using theoretical derived molecular descriptors, Environ. Toxicol. Chem., 9, 417-428 (1990)

[19] Mekenyan O.G., Ankley G.T., Veith G.V., and Call D.J., QSARs for photo-induced toxicity: acute lethality of polycyclic aromatic hydrocarbons to Daphnia magna, Chemosphere 28, 567-582 (1994)

[20] Mekenyan O.G., Ankley G.T., Veith G.V., Call D.J., QSAR estimates of excited states and photo-induced acute toxicity of PAHs, SAR and QSAR Environ. Research, in press (1994)

[21] Veith G.V., Mekenyan O.G., Ankley G.T., Call D.G., A QSAR of substituent effects on the photo-induced acute toxicity of PAHs, Chemosphere, 30, 2129-2142 (1995)

[22] Mekenyan O.G., Ankley G.T., Veith G.D., Call D.J., QSAR Evaluation of α-Terthienyl Phototoxicity, Environ. Sci. Technol., 29,1267-1272 (1995)

[23] Mekenyan O.G., Schultz T.-W., Veith G., Kamenska, O., Dynamic QSAR For Semicarbazide-Induced Mortality in Frog Embryo, J. Agr. Chem., in press (1994)

[24] Mekenyan O., Ankley G., Veith G., Call D., QSAR Evaluation of Binding of Aromatic Xenobiotics to the Aryl Hydrocarbon Receptor., Environ. Health Perspec., in press (1995)

[25] Mekenyan O.G., Veith, G.V., QSAR - Synergism Between Modeling and Physico Chemical Principles, SAR and QSAR Environ. Research, in press (1994)

REGULARITY AND SIMILARITY: PREDICTION TOOLS FOR METALLIC COMPOUNDS

Pierre VILLARS

Materials Phases Data System, Postal Box 1, CH-6354 Vitznau, Switzerland

ABSTRACT

The fundamentals of the constitution of an alloying system are determined by the crystal structure of its metallic compounds and its phase diagrams. Even if it is not understood there exists an obvious relation between the crystal structure of a metallic compound and its intrinsic physical properties. For example to date there are over 3.000 different crystal structures experimentally known, but high-T_c superconductors are only found to crystallize in one of a dozen different crystal structures. Analogous observations hold for all intrinsic physical properties.

RESUME

Les données fondamentales de la constitution d'un système d'un alliage sont déterminées par la structure cristalline des composantes métalliques et de son diagramme de phase. Même si l'on n'en comprend pas les raisons, il existe des relations évidentes entre la structure cristalline d'un composé métallique et ses propriétés physiques intrinsèques. Jusqu'à ce jour, plus de 3000 structures cristallines différentes sont connues expérimentalement mais les superconducteurs Haut-T_c ne cristallisent que dans une douzaine de cristaux différents. Il est de même, au sens restrictif, pour toutes les propriétés physiques intrinsèques.

1 INTRODUCTION

Knowing these fundamentals and these intrinsic physical property-crystal structure relations enables scientists to solve many problems in materials science, and therefore it is important to have easy access to these experimentally determined data, especially as the experimental work to determine such information is very time- and cost-intensive.

Recently, scientists working in this field have had the advantage of access to comprehensive, up-to-date handbooks for crystal structures as well as for phase diagrams [1, 4]. For intrinsic physical properties there exists no comprehensive handbook, but for some intrinsic physical properties, e.g. ferroelectrica and magnetic materials there exist some contributions [5, 6]. For crystal data there exists the database CRYSTMET (7). Very recently we started to plan a long-term project, the creation of the *PAULING FILE DATABASE,* a fully relational database which will contain crystallographic and 'key' intrinsic physical property data (numerical and factual) as well as phase diagrams (images) for all known *'non-organic' materials (systems)* covering the literature from 1900 to date. Under 'nonorganic' materials we include alloys, intermetallics, ceramics, minerals and inorganics. There are over 200.000 entries containing numerical and factual data, as well as over 25.000 entries containing image data expected.

Looking at the research activities of the last 20 years with respect to the number of systems investigated, grouped in binary, ternary, and quaternary systems, a shift in the ongoing research from binary to ternary and quaternary systems is clearly seen. The research field of multicomponent systems is a huge field for novel materials with optimized known intrinsic physical properties, e.g. tP68 $BFe_{14}Nd_2$ type permanent magnets, oP14 $Ba_2Cu_3O_7Y$ and cF492 $C_{60}K_3$ types high—T_c superconductors as well as completely new intrinsic physical properties, as the recent past has proven with the discovery of quasi-crystals. All this was achieved by moving from the binary to ternary and quaternary systems, but this trend confronts us with many additional 'difficult-to-handle' problems.

The experimental variables we have are a selection of the chemical elements, their possible combinations, their concentrations, the temperature and the pressure. Of enormous consequence from the practical point of view is the variety of combination possibilities of the chemical elements and the increase in the possible number of combinations on going from binaries to multinaries accompanied by a much larger available concentration range. Unfortunately, the number of available experimentally determined data is in inverse proportion to these opportunities. We have a relatively robust database for binaries, but a sparse database for ternaries, and almost no data for quaternaries (see Table 1).

Compounds	**Experimentally Known Systems**	**Maximum Number**
Unaries	100	100
Binaries	4,000 (81%)	4,950
Ternaries	8,000 (5%)	161,700
Quaternaries	1,000 (<0.1%)	3,921,225

Table 1. Experimentally known systems, relative to the estimated maximum number of compounds grouped by elemental composition. The starting number of chemical elements is 100.

Without very good guidelines, it is hopeless to search systematically for novel materials with an adequate success rate in multicomponent systems. Therefore at the moment, the only practical way is, with the help of the robust binary database, to find regularities and similarities valid in binaries, and then to extend these to the ternaries and quaternaries, in such a way that the binaries and others can be treated together; otherwise the regularities and similarities would be based on too few data sets and therefore would not be trustworthy. In addition, the regularities and similarities should show an accuracy clearly above randomness to be of practical use (these can easily be checked with the available determined data).

To increase the efficiency in the successful search for new metallic compounds, the main efforts should go towards creating an internationally accessible information-knowledge system linking the proposed *PAULING FILE DATABASE* (critically evaluated experimentally determined facts) with generally valid 'highest quality' regularities and similarities representing the *KNOWLEDGE BASE.* The problems involved in such a project have been reviewed by Westbrook [8]. The combination of the experience and intuition of the experimentalist together with access to the envisioned 'easy to use' information-knowledge system would very much help to coordinate world research activities. Furthermore, it would reduce the number of unwanted duplications as well as increase the probability of investigating the most promising systems first and not, as in the past, provide us with the systems in a random statistical sequence.

2 STRATEGY TO FIND FACTORS GOVERNING CRYSTAL STRUCTURES, INTRINSIC PHYSICAL PROPERTIES AND CONSTITUTION PROPERTIES

When we talk about factors governing either crystal structures, intrinsic physical properties or constitution properties we intend to reduce these to atomic property expressions of *their constituent chemical elements, so* that the existing experimentally determined data of their metallic compounds or systems *can be systematized.* For this purpose we consider only regularities and similarities as being of real practical value and able to systematize a large group of data with an accuracy in the range of at least 95 %. For the not yet experimentally investigated metallic compounds or systems, one proceeds in exactly the reverse way, assuming the validity of the regularity and similarity found for a well-defined group of data. The more data considered, the more trustworthy is a prediction based on such regularities and similarities. To make predictions one starts from the *chemical elements, their concentrations, and the tabulated atomic properties of their constituent chemical elements,* and one calculates the atomic property expression for the metallic compound or system concerned. Therefore, tabulated atomic properties are only of practical interest where they are known for most chemical elements with adequate accuracy. In principle, it would be sufficient to use as input only the atomic numbers and the compositions of the metallic compound of the considered system. Although from the theory the interactions in metallic compounds are well understood, it is not an easy task to evaluate the total energy of solids. As the energy of an isolated atom is on the order of about 10 GeV, while the cohesive energy responsible for the bonding situation (and therefore for its crystal structure and its intrinsic physical properties) is only on the order of 1-10 eV/atom, one must have a method that is accurate to one part of 10G, or better. The other fact that greatly complicates evaluating cohesive energy by theoretical methods is the number of particles involved. Given that a macroscopic solid may contain 10^{23} nuclei and electrons, it is impossible to determine the total energy of the crystal structure without some approximations.

Therefore it is sensible to adopt semi-empirical approaches based on experimentally determined data to search for the most reliable regularities and similarities. Analogous to first principle calculations one will have to see how far one can go with the simplification (approximation), without loosing the desired information, e.g. for the prediction of the crystal structure of AB metallic compounds by separating AB compounds with different crystal structures into distinct domains in an atomic property expression versus atomic property expression plot. If the approximations are driven too far one will loose the separation into different domains. As at the moment it is impossible to start only from the atomic property atomic number, it is obvious to try to find out which other atomic properties of the chemical elements are needed to describe the alloying behavior. Villars [9] concluded a survey of 53 different atomic properties as a function of the atomic number (182 variables in all when the different methods of determination are taken into account), and it was found that there were only five main groups, here called factors. The results are best seen in an atomic property versus atomic number plot as shown in Figure 1 for the pseudo-potential radii after Zunger [10] for s, p, and d elements.

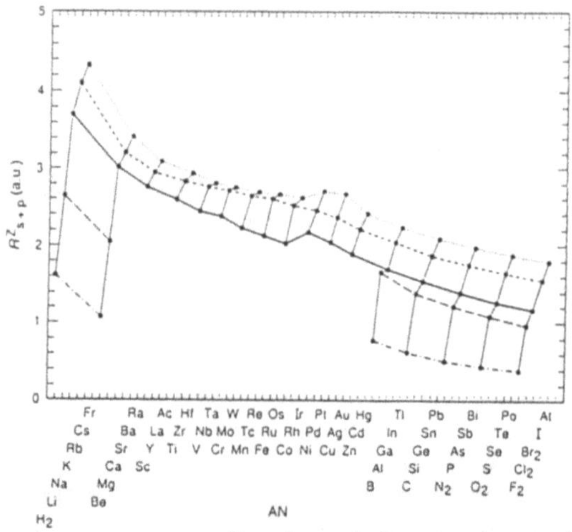

Figure 1. Atomic property versus a function of the atomic number plot of the pseudo-potential radii after Zunger [10]. In such diagrams the chemical elements are ordered by increasing group numbers and each group is ordered by increasing quantum numbers. This special atomic number scale was used instead of a linearly increasing atomic number scale because it gives units of comparable sizes for distances between atoms within any group as well as along any period. The scale chosen for the comparison, of course, does not affect the number of factors, only the appearance of the patterns changes.

In all but a few cases (19 out of 182) very regular symmetrical patterns were obtained in such plots: this means a regular behavior in a group with increasing quantum numbers as well as in a period with increasing atomic numbers.

It should be stressed that the equivalence of atomic properties belonging to the same group is of a qualitative, not a quantitative, nature. Adherence to one or another of the five factors is very obvious. Figure 2 shows three atomic properties belonging to the same factors plotted against each other, the cohesive energy E_c and the self-diffusion coefficient Q versus the melting point T_m of the chemical elements. As a first approximation linear dependencies may be seen to exist.

The five factors (groups) are:

- Size factor
- Atomic number factor
- Cohesive-energy factor
- Electro-chemical factor
- Valence-electron factor

An atomic property for each factor, for which accurate and complete data are available as well as (and this is very important for practical use) values that are independent of the constituent chemical elements of a compound or system are chosen. These are:

- Pseudo-potential radius after Zunger [10]
- Atomic number
- Melting point
- Electro-negativity after Martynov and Batsanov [11]
- Group number

By comparing these values with each other it is clearly seen that there exists no overall tendencies between them; they thus represent the most independent factors.

3 MATERIALS DESIGN BASED ON REGULARITY AND SIMILARITY

There exist three principal different ways of materials design based on regularity and similarity:

a) Extended 'one by one shot' approach

b) Constituent atoms towards materials approach

c) Materials towards constituent atoms approach

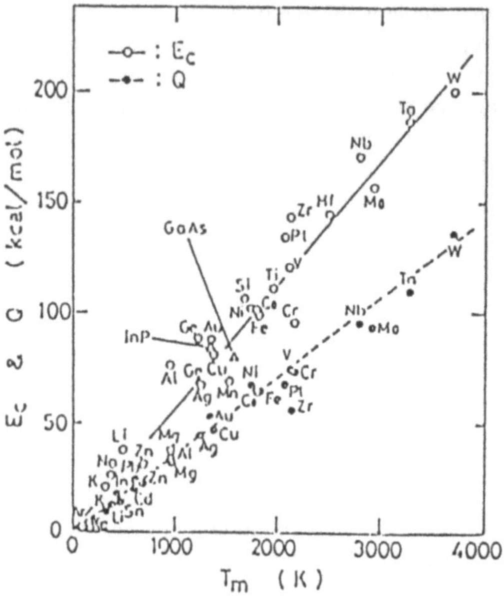

Figure 2. The atomic properties cohesive energy E_c and the self-diffusion coefficient Q are plotted versus the atomic property melting point T_m. All three atomic properties belong to the cohesive-energy factor.

The extended 'one by one shot' approach is the conventional one based on the use of concepts and behaviors in a more or less systematic way. Figure 3 shows the most outstanding concepts and behaviors known in materials science [12].

If we try to analyze the conditions that made it possible to find such concepts and behaviors we come to the following. It was not a new technique, a unique experimental observation or an abstruse theory which formed the take-off point. Rather it was the amassing of a critical volume of experimentally determined data in the literature permitting an individual with great insight to perceive an underlying pattern not previously apparent. The constituent atoms towards materials approach as well as the reverse materials towards constituent atoms approach represent new tools for materials design and follow the main steps listed below (Fig. 4):

- Prediction of compound formation within a system
- Prediction of formula in compound forming system

- Prediction of its crystal structures
- Prediction of its intrinsic physical properties
- Prediction of the influence of its morphological properties

The availability of the large amount of experimentally determined fundamental data like crystallographic data, intrinsic physical property data and phase diagram (constitution) data; *"DATABASE of the materials"* and the availability of atomic property data of the chemical elements; *"DATABASE of the chemical elements"* and *"All distinct different techniques to re-arrange materials data"* from a statistic to an ordered distribution in parameterized atomic property representation in linking the above mentioned databases in a strategic way, would enable us to systematically build a *KNOWLEDGE BASE by* finding all existing regularities and similarities.

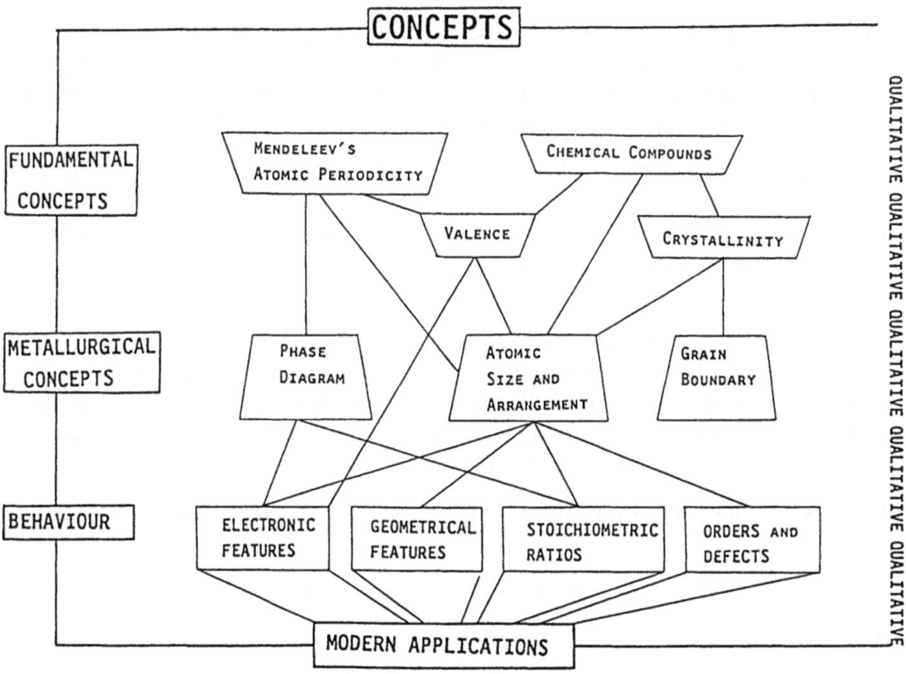

Figure 3. Chart showing the linkage of key concepts of metallic compounds formation and behavior (rectangles) from metallurgical concepts and ultimately from fundamental concepts of chemistry and physics.

Of course this represents the ideal case, and we would assume that the experimentally determined data are correct and we would have the knowledge of all existing distinct different techniques to re-arrange data. Unfortunately this is not reality, but nevertheless it shows us the boundary conditions we have to watch carefully.

Concerning the quality of both databases (covering materials data and atomic property data) there exist substantial discrepancies between data-sets determined by different scientists. I experimented by comparing crystallographic data and phase diagram data of materials in preparing 20.000 handbook pages [1, 2] covering over 35.000 publications and found that there exist over 30 % major discrepancies. In preparing the Structure Type Atlas [3] and the atomic environment classification in collaboration with Jo Daams and Jan van Vucht [13, 15] we observed that 25 % of all published crystal structures have improbable interatomic distances within the crystal structure. This rate of inaccuracy is high, since the results of the regularities and similarities are hidden, when critically evaluated data are not used. It is highly recommended to create a sub-database containing only independently verified 'highest quality' data of materials. The situation for atomic property data is analogous.

To search for all existing distinct different techniques we intend to analyze all published regularities and similarities in world literature. This approach will certainly not guarantee completeness but will provide a good start. Below, I list three different techniques, each successful in solving a specific materials problem:

1) Atomic property of chemical element A versus atomic property of chemical element B - crystal structure of materials AB plot. The AB metallic compounds are nicely separated into different structure type domains in an atomic number of the constituent chemical element A versus atomic number of the constituent chemical element B plot [16] (Fig. 6).

2) *Atomic property—expression$_1$ (A,B,C) versus atomic property—expression$_2$ (A,B,C) versus atomic property—expression$_3$ (A,B,C) - constitution property of A-B-C systems.* The quantum formation diagram (QFD) is very successful in separating ternary A-B-C systems in two volumes, the compound-formation and the absence of compound-formation volumes (Fig. 5) [17].

For the QFD the following three expressions are used in the parameterized representation:

$$\left(\left|\Delta R_{AB}\right| + \left|\Delta R_{AC}\right| + \left|\Delta R_{BC}\right|\right) / 3$$

$$\left(T_A / T_B + T_A / T_C + T_B / T_C\right) / 3$$

$$\left(\left|\Delta \mathrm{V}_{\mathrm{AB}}\right| + \left|\Delta \mathrm{V}_{\mathrm{AC}}\right| + \left|\Delta \mathrm{V}_{\mathrm{BC}}\right|\right) / 3$$

where $T_A > T_B > T_C$

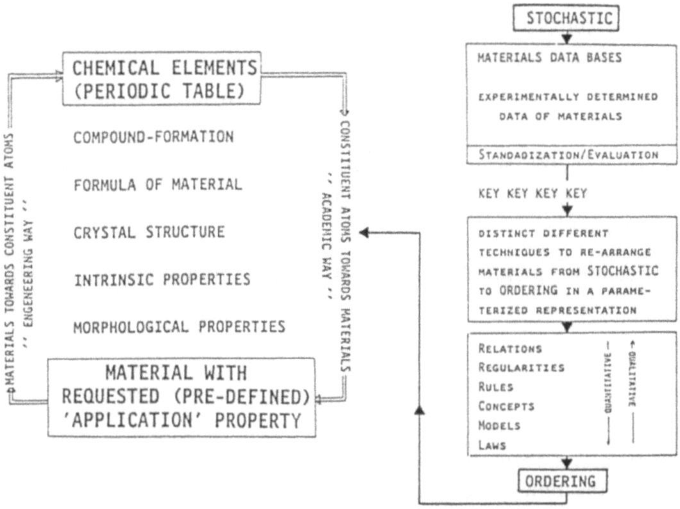

Figure 4. Chart showing the connections between databases and materials towards constituent atoms and constituent atoms towards materials

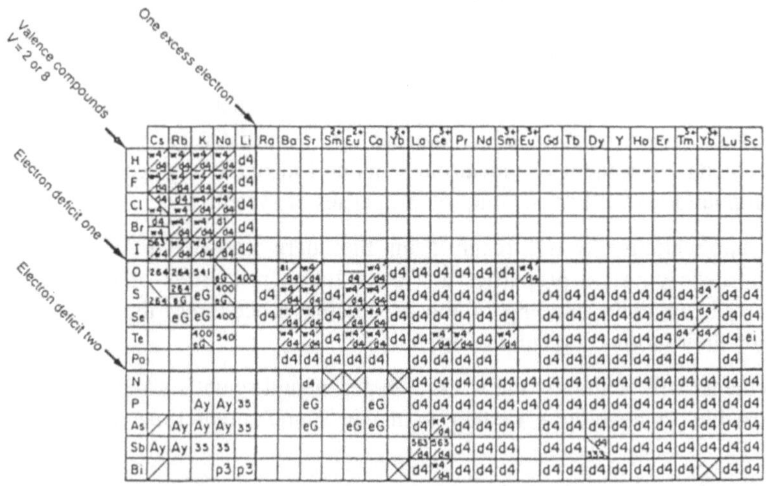

Figure 5. Three-dimensional compound-formation plot of the section for valence electron= 2 for ternary systems. Each symbol represents one ternary system. The triangle stands for no compound formation determined by experiment; a circle stands for no compound formation by extrapolation from established experimentally determined binary phase diagrams, all showing absence of compounds, and a black circle stands for experimentally determined compound forming systems.

R stands for the pseudo-potential radii after Zunger [10], T for the melting points and V for the valence electrons (group number) of the chemical elements A, B, C respectively.

3) Space X,Y,Z versus symmetry versus geometrical restraints. With the help of the space group theory and geometrical restaints observed to be valid for all experimentally known crystal structures one can generate all possible combinations of point-sets within a certain space group and exclude all those which lay outside the geometrical restraints [18].

For each of the three afore-mentioned materials design approaches we give a practical example.

CONCLUSION

The crystal structure, intrinsic physical properties and constitution properties of metallic compounds (systems) are, at the present level of our understanding, only partially predictable from fundamental atomic properties of their constituent chemical elements. We therefore discuss the development of a sophisticated, computerized information knowledge system for materials design science.

This system should bring into one computer environment the enormous amount of experimentally determined data with the *PAULING FILE DATABASE* and known regularities and similarities with the *KNOWLEDGE BASE.*

This system could be used to test the validity of already published work, and to search for all existing regularities and similarities in quantitatively atomic property expressions of its constituent chemical elements and intrinsic physical properties of metallic compounds as well as constitution properties of systems.

Successful applications do not result from the mere knowledge of the crystal structure, intrinsic physical properties and constitution properties of materials, but require as well, an understanding of the processing procedures; these are mainly influenced by their morphological properties.

Such refinements will be mainly the work of experimental scientists and it will hardly ever be possible to develop a system to make 'sound' predictions of processing techniques.

The role of regularities and similarities in materials science can be best summed up in the words of F. Crick [19] on the role that theorists play in exploring the complex world of nature:

"The best a theorist can hope to do is to point an experimentalist in the right direction, and this is best done by suggesting what directions to avoid".

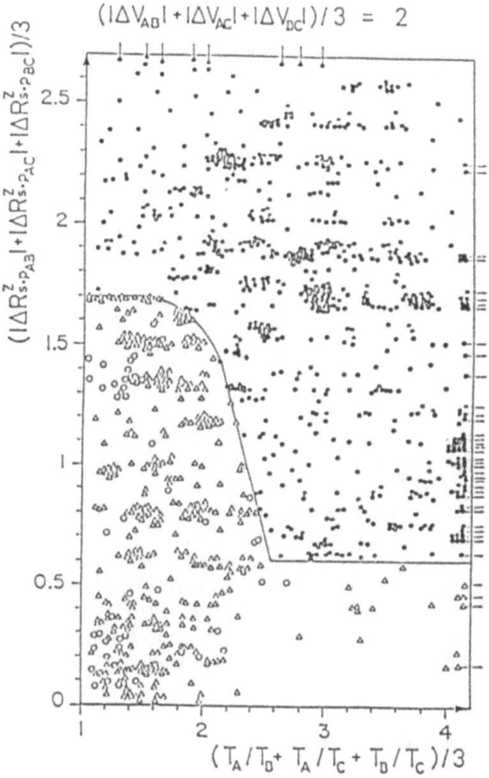

Figure 6. Example of an atomic property of the chemical element A versus the atomic property of the chemical element B - crystal structure of materials AB plot. A Mendeleev-type arrangement of the atomic number was used as atomic property. For polymorphic AB compounds we distinguished high-temperature/low-temperature modifications by indicating them in the upper right/lower left corner, and high-pressure/low-pressure modifications in the upper left/lower right corner, respectively. In cases where information is lacking or where both temperature and pressure are involved, we separated the entries for the two modifications ml and m2 by a horizontal line. Structure types are: 'eG' hP12 NaO, 'Ay' oP16 NaP, 'ei' hP4 AsNi, 'd4' cF8 ClNa, 'we' cP2 ClCs, 'dl' oP8 GeS. Numbers appear as symbols for crystal structures where the atomic environment was not investigated.

KEYWORDS

Pauling file database, knowledge database, regularity, similarity, morphological properties, crystal structure, metallic compounds, analogous observation

REFERENCES

[1] Villars P., Calvert L. D., Pearson's Handbook of Crystallographic Data for Intermetallic Phases, 2nd edn., 1-4. ASM International, Materials Park, OH (1991)

[2] Villars P., Prince A., Okamoto, H., Handbook of Ternary Alloy Phase Diagrams, 1-10. ASM International, Materials Park, OH (1995)

[3] Daams J.L.C., Villars P., van Vucht J.H.N., Atlas of Crystal Structures for Intermetallic Phases, 1-4. ASM International, Materials Park, OH (1991)

[4] Massalski T.B., Okamoto H., Subramanian P.R., Kacprzak L., Binary Alloy Phase Diagrams, 2nd edn., 1-3. ASM International, Materials Park, OH (1991)

[5] Landhold-Börstein, Ferroelectrics and Related Substances, III/16 subvolume b., Springer-Verlag, Heidelberg (1982)

[6] Landhold-Börstein, Magnetic Properties of Metals, III/19 subvolumes b,c,dl,d2. Springer-Verlag, Heidelberg (1991)

[7] CRYSTMET CAN/SND, Canadian Institute for Scientific and Technical Information, NRCC, Ottawa, Canada (1995)

[8] Westbrook J. H., MRS Bulletin, February, 44 (1993)

[9] Villars P., Journal of the Less-Common Metals, 92, 215 (1983)

[10] Zunger A., In Structure and Bonding in Crystals, O'Keeffe M., Navrotsky A. (Eds) Academic Press, New York (1981)

[11] Martynov A.J., Batsanov S. S., Russian Journal of Chemistry, 25, 1737 (1980)

[12] Westbrook J.H., Metallurgical Transactions A, 8A, 1327 (1977)

[13] Daams J.L.C., Villars P., van Vucht J.H.N., Journal of Alloys and Compounds, 182, 1 (1993)

[14] Daams J.L.C., Villars P., Journal of Alloys and Compounds, 197, 243 (1993)

[15] Daams J.L.C., Villars P., Journal of Alloys and Compounds, 215, 1 (1994)

[16] Villars P., Mathis K., Hulliger F., In The Structures and Binary Compounds (Cohesion and Structures, de Boer F.R., Pettifor D.G. (Eds) 2 North-Holland, Amsterdam (1989)

[17] Villars P., Journal of the Less-Common Metals, 109, 93 (1985)

[18] Villars P., Dissertation N°. 6787, ETH Zürich, Switzerland (1991)

[19] Crick F., In What Mad Pursuit: A Personal View of Scientific Discovery. Basic Books, New York (1988)

Part Two

New Data Channels: Industrial Information Tools

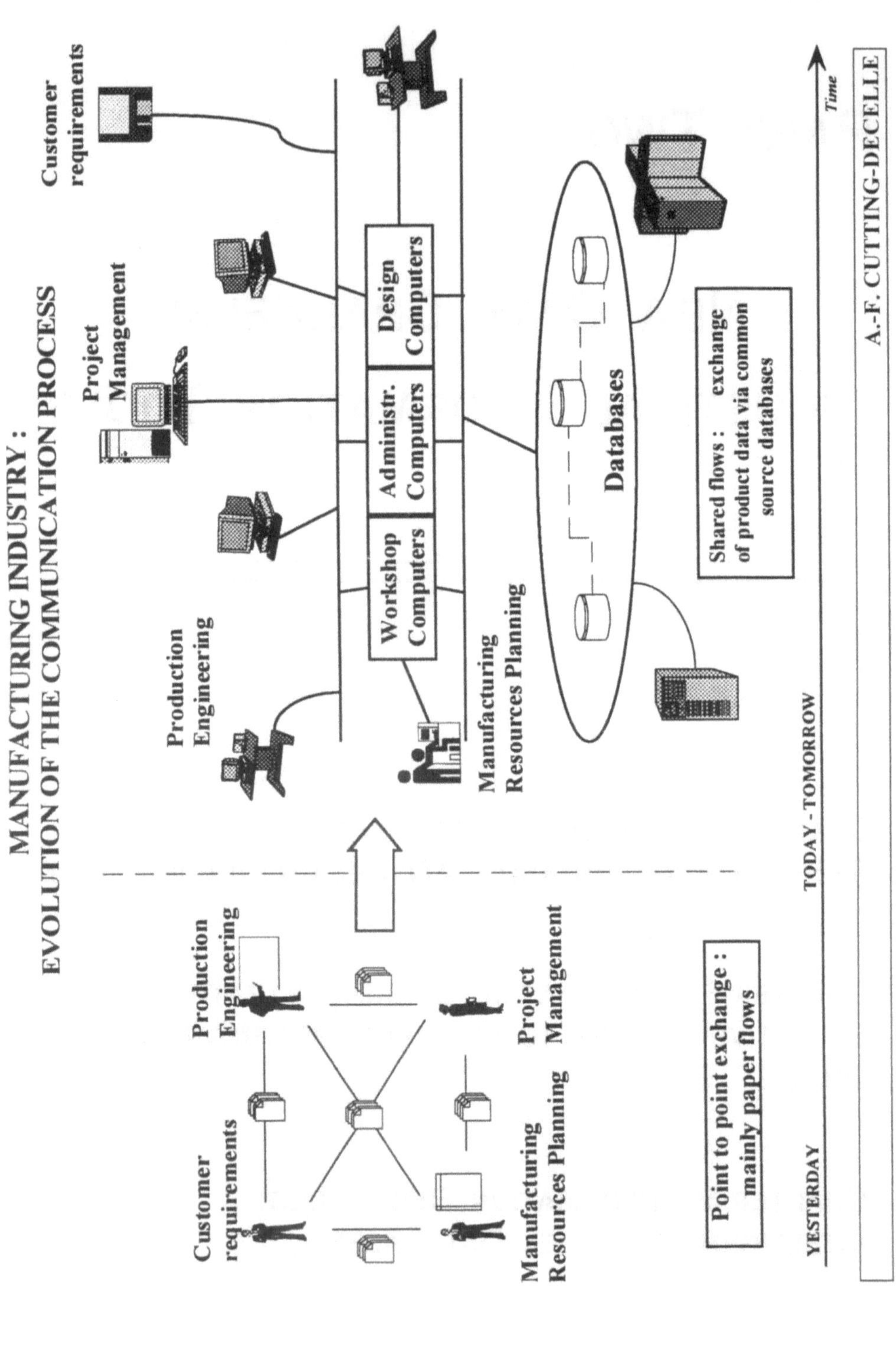
MANUFACTURING INDUSTRY :
EVOLUTION OF THE COMMUNICATION PROCESS
Customer requirements
Production Engineering
Project Management
Manufacturing Resources Planning
Point to point exchange : mainly paper flows
YESTERDAY
TODAY - TOMORROW
Customer requirements
Production Engineering
Project Management
Manufacturing Resources Planning
Workshop Computers
Administr. Computers
Design Computers
Databases
Shared flows : exchange of product data via common source databases
Time
A.-F. CUTTING-DECELLE

Chapter 4

INDUSTRIAL INFORMATION EXCHANGE: STEP, STANDARDS AND CAD DESIGN NEEDS

STEP: THE ISO 10303 STANDARD FOR PRODUCT DATA EXCHANGE AND REPRESENTATION

Alain BEZOS

Association GOSET, 107-111 Avenue Georges Clemenceau, 92000 Nanterre, France

ABSTRACT

This presentation describes the international standardization project, STEP, whose final goal is to develop a mechanism enabling the capture of any technical data related to industrial products throughout their life cycle. The completeness of such a representation will make it usable not only for neutral file exchange between CAD systems, but also as a fundamental support for the design and sharing of product databases and for archiving data. After about ten years, the STEP project has already produced the first parts of an International Standard for the exchange of product data. But its main achievement is the definition of a new technological approach for the representation and the use of product data, well fitted to the new data management techniques, and internationally accepted.

RESUME

Cette étude a pour but de présenter le projet international de normalisation STEP, dont l'objectif ultime est de développer un mécanisme permettant le traitement de toutes les données techniques relatives aux produits industriels tout au long de leur cycle de vie. La plénitude d'une telle représentation la rend utilisable non seulement pour l'échange de fichiers neutres entre des systèmes de CAO, mais en fait aussi un support fondamental pour la conception et le partage de bases de données de produits, ainsi que pour l'archivage. Après environ dix ans, le projet STEP a déjà produit les premières parties d'une Norme Internationale sur l'échange de données de produits. Cependant sa principale réalisation concerne la définition d'une nouvelle approche technologique relative à la représentation et à l'utilisation de données de produits, bien adaptée aux nouvelles techniques de gestion des données, et reconnue de façon internationale.

1 INTRODUCTION

Starting with 50 people to replace IGES, SET and VDA, they were 300 ten years later, standardizing a huge data structure able to contain all product data, to serve as a basis for the development of integrated application software and to keep and manage the enterprise memory and knowledge.

2 THE INDUSTRIAL PROBLEM

Today the availability and the right use of information have become absolutely necessary for the success of an industrial enterprise.

Various implemented computer aided systems (CAX) generate, process and manage data abundantly in all departments of the enterprise as well as partners and sub-contractors of the enterprise. The availability of the right data at the right place and at the right time is a key factor for an effective process in design, manufacturing, maintenance and recycling, that is to say, for the industrial implementation of Concurrent Engineering.

Data describing a product and its environment to manufacture the product play an important part in manufacturing industries. Data describing a single product are generated and processed by many systems. The different system native formats and structures handicap the use of results from one process step to the next. It follows that data are generated repeatedly and managed redundantly.

Moreover, the life cycle of CAX systems is lower than the life cycle of products. The enterprise memory cannot be supported by these systems.

Today, product data do not belong to the product but to an enterprise division, a subcontractor, a partner, or even worse, to a given CAX system which is the only one able to understand and exploit them.

Due to the increase of partnership projects, to the necessity to promote and use concurrent engineering techniques in order to extend enterprise competitivity and product quality and to reduce delay on the market, the management and sharing of technical data is a critical problem for industry.

Product Data Breakdown

ENTERPRISE		
Design	Fabrication	Assembly
Data System A	Data System B	Data System C

S/CONTRACT.
Design
Data System H

PARTNER 1	
Design	Manufacturing
Data System D	Data System E

S/CONTRACTOR	
Simulation	Manufacturing
Data System F	Data System G

3 THE NEED FOR A STANDARD

Today, many CAX system suppliers propose software solutions integrated with a data base for solving the industrial problem of process integration through the sharing of data. However, these solutions today are still closed solutions which maintain the user under his supplier's control, impose work methods on him, do not allow him communication with partners variously equipped and do not ensure information permanence.

The 3 Axes of Product Data

Product Lifecycle	Product structure	Product data types
Destruction	Elementary component	2D geometry
Use	Product part	3D wireframe geometry
Maintenance	Assembly	3D surface geometry
Installation	System	3D solid geometry
Assembly	Group of systems	Features
Testing		Tolerances
Manufacturing		Drawings
Simulation		Materials
Delayed design		Product configuration
Design		Process data
Specifications		Simulation data
Procurement		

For a long time, users have defined standards to enable exchange of data between CAD systems. Whatever the success of these standards, such as IGES, SET or VDA, they are not internationally used and are restricted to data exchanges using physical files for the design phase and mainly geometrical data. We had to go further, i.e. to process all data present throughout the product life cycle to enable a more direct integration of applications while taking advantage of modern data processing methods such as formal language of data definition or "object oriented approach". Moreover, the specifications thus defined were to meet industrial requirements, i.e. that data models be developed from precise specifications established by users and representing a synthesis of their industrial know-how.

Much effort and research have been and will still be necessary to undertake this ambitious task.

ISO/TC184 decided in 1984 to create a sub-committee, the SC4, for developing such a standard, under the name STEP. At the same time, CEC/DGXIII launched a research project, the Esprit CAD*I project, on that same subject. After a few years of labor, in 1989, the first STEP version brought to a vote was unanimously rejected and even if the European project did achieve some interesting results, it brought no real industrial response.

The problem was thus more complicated than expected. As the continuation and the extrapolation of existing solutions were unsuccessful, a new start was made on new theoretical bases, with new tools, and a new SC4 organization. Concurrently, new research projects were launched around the world as well as many prototype development projects to evaluate the new theoretical concepts.

Some recent European projects :

- Esprit CADEX and PRODEX : Geometry and finite elements
- Esprit NEUTRABAS and MARITIME : Ship building
- Esprit PROCESSBASE : Process plants
- Esprit NIRO and INTERROB : Robotic application
- Esprit ATLAS and Euréka CIMSTEEL : BTP field
- Esprit PISA : General methodology and tools
- Euréka CIMSTEEL : Steel works

Today, we have the first results. The initial release of STEP has obtained a status of International Standard, thus demonstrating the project's credibility and maturity. This first version deals only with a limited field and must still prove itself industrially, but it has reliable and stable bases needed for carrying out work.

Content of the initial release of STEP :

- Part 1 : Overview and Fundamental Principles

- Part 11 : The EXPRESS Language
- Part 21 : Clear Text Encoding
- Part 31 : Conformance Testing: General Concepts
- Part 41 : Fundamentals of Product Description
- Part 42 : Geometric and Topological Representation
- Part 43 : Representation Structures
- Part 44 : Product Structure Configuration
- Part 46 : Visual Presentation
- Part 101 : Drafting
- Part 201 : Explicit Drafting
- Part 203 : Configuration Controlled Design

The first battle is won, but we must not lose the war. From now on, industrial users must help the STEP experts to specify their needs in the industrial field. Great efforts must also be made to check and rationalize the fantastic explosion of application protocol development and to implement strategic planning of future work.

Some developments in process :

- Part 22 : Standard Data Access Interface
- Part 45 : Materials
- Part 47 : Shape Tolerance
- Part 48 : Form Features
- Part 104 : Finite Element Analysis
- Part 105 : Kinematics
- Part 202 : Associative Drafting
- Part 204 : Mechanical Design using B Rep
- Part 205 : Mechanical Design using Surface Rep
- Part 207 : Sheet Metal Die Planning and Design
- Part 208 : Life Cycle Product Change Process
- Part 209 : Composite and Metallic Structures
- Part 210 : Printed Circuit Assembly Design
- Part 211 : Electronic Test Diagnostics
- Part 212 : Electrotechnical Plants

- Part 213 : Numerical Control for Machined Parts
- Part 214 : Automotive Design Process
- Part 215 : Ship Arrangement
- Part 216 : Ship Molded Forms
- Part 217 : Ship Piping
- Part 218 : Ship Structures
- Part 221 : Functional Data for Process Plants
- Part 222 : Composite Structures
- Part 225 : Structural Building Elements

4 THE STEP SPECIFICITY

The final international objective undertaken in STEP project is to develop a mechanism able to represent overall product definition data throughout its life cycle. The completeness of such a representation will be a suitable tool for the exchange and the archiving of neutral files between CAX systems as well as a fundamental support for the design and sharing of product data bases.

The STEP specificity is essentially the methodology used for its development which consists in the adoption of a three layered architecture, independently described with formal languages and thus separating the conceptual model of information, from the physical representation of data, and from the application views.

The logical layer contains the global conceptual model of product data. This model which is a set of particular models (geometry, kinematics, tolerances, draughting, etc...) describes in a non-ambiguous and non-redundant manner the overall information needed for product representation, i.e. independently from the use which will be made of it.

The integration phase of these particular models, which are named integrated resources, enables their proper imbrication, coherence and their adequate level of genericity.

A formal language of data definition, EXPRESS, was developed and used to meet these requirements.

The physical layer is made up of proper schemes for the STEP implementation. If today it concerns only the definition of a neutral file format for data exchange, which is obtained by direct translation of the EXPRESS language, other uses are being processed such as the definition of product data bases, data structures for CAX softwares and a standardized data access interface.

The application layer contains the various views needed for the use of the standard within a given application context. These views, named Application Protocols, include the following elements :

- An application activity model, AAM, i.e. a specification defining the use of the data and the boundaries of the application field, defined in natural language and formalized with IDEF0,
- An application reference model, ARM, i.e. an exhaustive analysis of the semantics conveyed by the application. This analysis is formalized in a diagram containing all manipulated data and their relationships which is expressed with a formal language such as NIAM or IDEF1X and in EXPRESS,
- An application interpreted model, AIM, which expresses the connection between the application reference model and the global conceptual diagram of the logical layer, i.e. specifying which entities and constructions have to be used and with which additional constraints and specializations. This model is written in the EXPRESS language.
- A set of test criteria, ATS, which will allow for checking and certifying the conformity of the implementations of the application protocol with the standard.

STEP Architecture

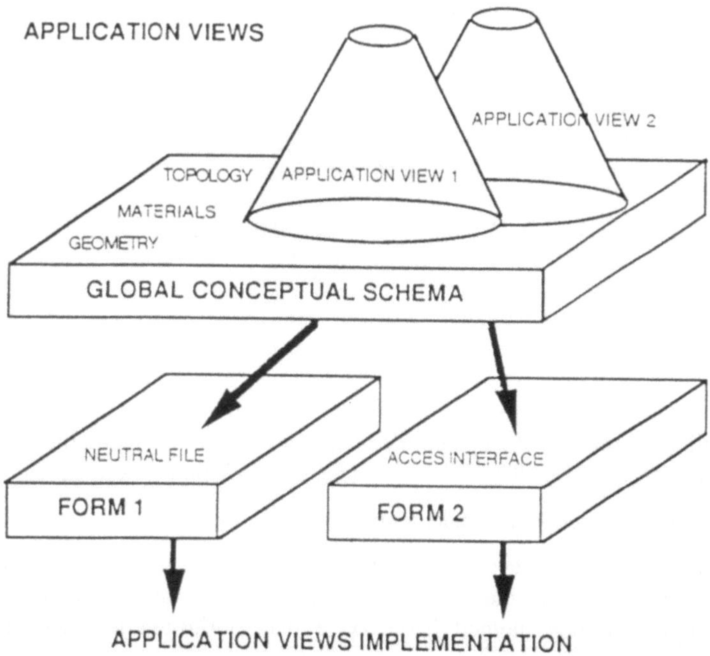

The architecture presented above is greatly simplified as compared with reality, even if it may already appear quite complicated. As an example, it covers up the difficulty in defining the global conceptual schema, i.e. to integrate in a single form concepts shared by many applications.

In particular, the application protocol interoperability problem, i.e. the possibility of using several different application protocols on the same data base is not solved today and will probably imply a new organization of the global conceptual schema and specific rules for the use of the corresponding models in protocols.

The complexity of manipulated notions and of their organization is one of the obstacles that STEP will have to avoid.

5 THE USE OF STEP

We can briefly retain the present state of data exchange practice. The existing national standards after great effort and important investments made by users and suppliers have reached a satisfying reliability degree. They enable, IGES for drafting, VDA for simple surfaces and SET for modeling complex surfaces and solids, the exchange of data with neutral files between CAD softwares dealing with related field. Subsisting difficulties referred to the various representations of the same concepts manipulated by the various CAX systems. These problems are currently solved with the development of assisting tools for exchange which are able to execute necessary transformations between related concepts.

Concept Representations

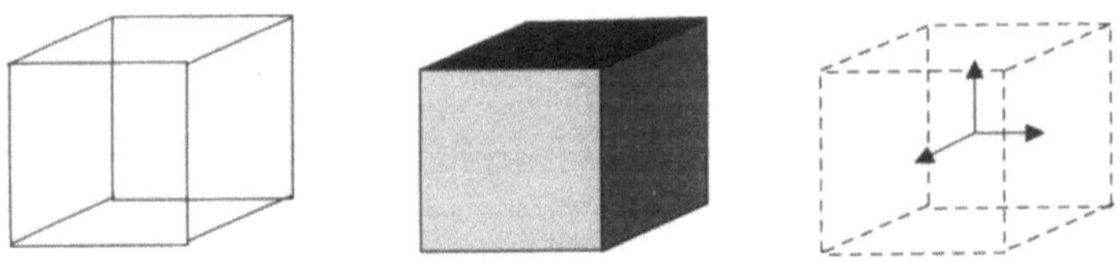

Wireframe Surfaces Primitive Solid

STEP will not directly solve this problem and the study of the available specifications shows this clearly. What then is the advantage of developing a new standard ?

The advantage is much more fundamental.

First of all, STEP is based on a modern technology and adapted to performing data processes allowed with the use of "object oriented" approach. The use of a specification definition method based on a data definition language, EXPRESS, allows a strict and consistent formalization of data models. Writing the specifications for using such a language to define not only product data but also the various constraints that must be respected by these data as well as the processing to be applied, makes these directly usable for establishing relational or object oriented data bases. Moreover the standardization of a Data Access Interface will specify access functions to a data structure in conformance with AP specifications, and thus enable applications standardization and interchangeability.

On the other hand, the application protocol concept enables users to retain the expertise of their profession. Indeed, they can define their requirements concerning product definition data to be represented, in the form of formalized specifications, using methods such as NIAM.

After that, the modeling experts will convert these specifications into EXPRESS data models based on STEP integrated resources. Then, these protocols could be nationally, regionally or internationally standardized according to their generality or remain proper specifications for an enterprise or a project.

This technology could be used for specifying data bases which are used for the development of many applications. These applications based on the same standardized product data model will be perfectly integrated. Thus STEP and its application protocols will not only be a method for exchange of data between similar softwares, but also a way to integrate successive applications enabling computerization of overall products life cycle, i.e. for data sharing.

Data Integration

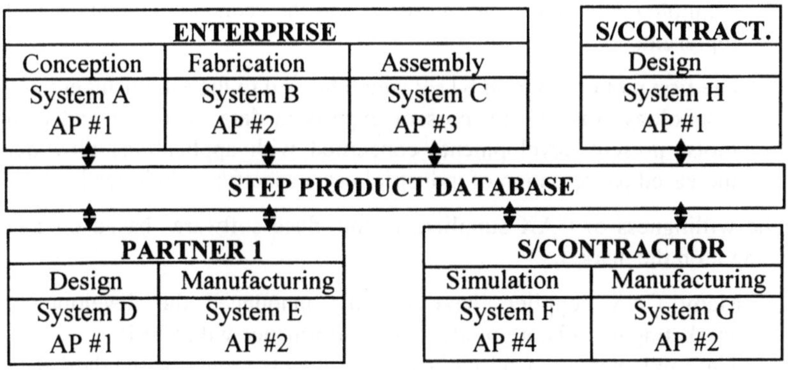

The STEP stake is to create an open technology which will allow the implementation of standardized technical data bases containing all product definition data needed for the management of its life cycle.

CONCLUSION

Today, we are not sure of bringing STEP to a successful conclusion, so much still remains to be achieved.

After ten years of work, the STEP project has not yet provided a standard for exchange able to replace IGES or SET, but the STEP project has led to the implementation of a new technological approach for product definition data representation.

The success of STEP as a necessary tool for the management and sharing of product data and for the simultaneous engineering and total quality concept implementation will in particular depend on the following factors :

- The implementation of a strategy within standardization organizations for carrying out standard developments,

  ISO/TC184/SC4 has created an advisory group responsible for strategic planning, the SPAG. This group, after having established a global strategy for the sub-committee has today to define application protocol priorities and their applicability in terms of covered field.

- The support and participation of industrial users to guarantee that the standards meet their requirements

  After consultation meetings with industrial users, CEC/DGIII created in 1991 an advisory group on product data technology, the PDTAG. One of this group's goals is to sensitize industrials for considering these technologies within their enterprise and the economic advantage that they represent.

  The first part of work on STEP which has led to the establishment of a new technology was the business of experts in data modeling and computer tools; present developments concerned with applications now need an increased participation of industrial users specifying their needs.

- The willingness of CAX suppliers to introduce softwares based on STEP or using it on the market,

  Many CAX suppliers have already announced the development and marketing of software products in conformance with STEP.
  The implementation of application protocols around a standard data base which will allow a share of product definition data between heterogeneous CAD/CAM systems has to be carried out. To that end, great projects have started, such as the ProSTEP project in the German automotive industry or the AeroSTEP project in the U.S.A., gathering Boeing and GE, P&W and RR motors as well as their suppliers, Dassault Systems, EDS and Computervision.

- The organization of a STEP expertise center network, working together for the quality of specifications, the promotion of their use, the certification of application softwares and help for industrial users.
- The EPDEN Network (European Product Data Exchange Network) has been set up since 1992, with the following organizations :

  CADDETC(U.K.), GOSET (France), IVF (Sweden), ProSTEP (Germany), SINTEF (Norway), TNO (Netherlands) and UNINOVA (Portugal). Moreover, many contacts and collaboration agreements exist with PDES, Inc. (U.S.A.) and the STEP Center (Japan).

  From now on, these centers must efficiently cooperate on industrial projects to validate the STEP concepts and exchange their experiences for the uniformity of the standard interpretation and implementation.
- The implementation of the transition to STEP which will enable industrial users to gradually and efficiently migrate from the use of existing national standards to the use of STEP.

  Such projects are now launched and researched. The goal for industry is not to lose the investment made on existing standards.

  From a technical point of view, it will be necessary to develop adapted application protocols and translation software tools and to develop and implement specific methodologies for the use of STEP.

Considering the necessity to manage the growing complexity of the international standard STEP, as much for its development as for its implementation and use, it appears that only a concerted and wilful effort by standardization organizations, public authorities and industrial companies will ensure the success of STEP.

KEYWORDS

STEP, data, product, data exchange, modeling, international standard, ISO, CIM, CAD.

REFERENCES

[1] Palmer M., Issues and recommendations for a STEP Application Protocol Framework, ISO/TC184/SC4/SPAG/N81, US Department of Commerce - NISTIR 4755, Jan.(1994)

[2] Interim report, ANSI/X3/SPARC study group on database management system, ACM SIGMOD, vol 7, ní 2 (1975)

[3] Bezos A., La technologie STEP pour la représentation des données deproduit, Normatique, n° 27, May (1991)

[4] Cutting-Decelle A.-F., Dubois A.-M., Product data handled by STEP: Considerations and proposals, Revue Internationale de CFAO et d'Infographie, Special Issue, vol 9, n° 3, (1994)

[5] Danner W., Stanford D., Yang Y., STEP Resources Integration: Semantic and syntactic rules, NISTIR 4528, March (1991)

[6] Owen J., Woodall A., STEP application protocols: Why and How?,ISO/TC184/SC4 WG1 N465, January 19, (1990)

[7] Arbouy S., Bezos A., Cutting-Decelle A.-F., Diakonoff P. , Germain-Lacour P. , Letouzey J.-P., Viel C. , STEP: Concepts fondamentaux, AFNOR (1994).

MATERIALS PROPERTY DATA IN STEP

Norman SWINDELLS *

Ferroday Ltd, 14 Mere Farm Road, Birkenhead L43 9TT UK

ABSTRACT

Products possess properties, in addition to their shape and dimensions, which are often associated with the materials from which the product is made. Part 45 of the ISO Standard 10303 provides the means to describe these properties, and the conditions under which they are valid. The development of this Part means that STEP can now describe tangible products and not only their shape representation. It could also be used to describe any measurable property of a product which is not part of the shape, such as the aerodynamic properties of a wing. This Part 45 has been developed in response to the requirements of the materials engineering community. The following presentation is intended to describe the main features of this Part, along with its objectives.

RESUME

En plus de leur forme et de leurs dimensions, les produits possèdent des propriétés souvent en relation avec les matériaux dont ils sont constitués. La Partie 45 de la norme internationale ISO 10303 fournit les moyens de décrire ces propriétés, et les conditions de leur validité. Le développement de cette Partie 45 a pour conséquence de permettre à STEP de décrire des produits "tangibles", et pas seulement leur forme. Elle peut aussi être utilisée pour décrire n'importe quelle propriété mesurable d'un produit autre que sa forme, comme par exemple les propriétés aérodynamiques d'une aile. Cette Partie 45 a été développée en réponse aux besoins de la communauté des ingénieurs spécialisés dans l'étude des matériaux. La présentation qui suit a pour but de décrire les principales caractéristiques de cette Partie ainsi que ses objectifs.

1 INTRODUCTION

Engineering materials are an integral part of the success of modern engineering, and information on materials is essential for the effective creation of new products and for anticipating their behavior throughout their lifecycle. However, the current situation

---

* Dr. N. Swindells, CEng, FIM, is the Managing Director of Ferroday Ltd, 14 Mere Farm Road, UK-Birkenhead L43 9TT and a member of the Materials Task Group of the ISO TC184/SC4/WG3.

with data from engineering materials is equivalent to the time before screw threads were standardized. Every manufacturer of materials provides information to his customers in a unique manner and their main aim is to provide this information in a form which can be understood by other people. Increasingly, information has to be understood by other computers and materials information is not in a suitable state, in general, for this to be achieved without a great deal of effort.

The main creators and suppliers of engineering data on materials are the producers of the materials and they expend a great deal of time and expense in measuring the data values and distributing them. The major limitation on the use of new materials and processes by design and production engineers is however still the availability of reliable data in a form that they can use directly in their computer-aided engineering (CAE) systems, and the supply of this information has not kept pace with the development of the computer systems on which manufacturing industry now relies. In order to achieve this connection between the materials producers and their engineering customers, the materials data will itself have to be engineered for the purpose of using it in CAE systems.

2 THE ENGINEERING OF MATERIALS INFORMATION

An engineering approach to the creation and management of materials information is achieved by the use of specifications for the information based on the use of standards, just as with any other engineered product, and there are several important developments which are directed to achieving this goal.

ASTM E-49

In the USA, a Committee, E-49, of the American Society for Testing Materials (ASTM) has been working for several years on this problem and the first standards are now starting to emerge [1]. Their attention has been concentrated on establishing standards for the recording and presentation of results from ASTM test methods for materials.

Some European Developments

The National Physical Laboratory in the UK has taken the lead, in collaboration with the British Plastics Federation, to establish standards for the recording and presentation of some single and multi-point data on plastics. The BSI standard (BS7008 Part 1) for single point data has now been proposed for adoption by ISO[2] and this work is now being extended to multi-point data in co-operation with German plastics manufacturers.

The AFNOR Standard Z99.01 has been used successfully by Vinard et al [3] to provide a standardized structure and meaning for materials data so that it can be transferred between different sites and between producers and customers who use the standard.

Within the European Union (EU) as a whole however the situation is confused. Several ad hoc initiatives in particular sectors of the materials producing industry have taken the lead in engineering their catalogue information into a usable and standardized form, which makes it potentially more valuable. The pioneers have been the CAMPUS® consortium of plastics manufacturers [4] and the European Aluminium Association of aluminium manufacturers. Other groups of manufacturers of engineering materials, such as copper, magnesium, cast iron, have responded to these actions with initiatives of their own.

The results of these ad-hoc initiatives satisfy the requirements of the *producers* of material products but mean that the *users* still see a variety of formats and incompatibility between the different material sectors for the *input* of data into their engineering systems. Also the scope of these agreements is limited to a restricted set of the features of a product and they are not sufficient for the full capability of CAD, CAE and CIM systems.

The ISO Standard for the Exchange of Product Data (STEP) is being created in a collaborative effort by materials users to overcome these problems for all manufactured products, including materials, on a global scale.

3 INTRODUCTION TO STEP

The new ISO Standard for the Exchange of Product Data (ISO 10303), known as STEP, is being developed to provide the global manufacturing industry with a single standard for the computerized representation and exchange of data for all manufactured products. The aims of the standard are to provide a complete, unambiguous, computer-understandable description of a manufactured product throughout its life cycle. The nature of this description makes it suitable for neutral file exchange between different computer systems and as a basis for implementing and sharing databases and for archiving product data [5].

To enable data created by one computer program to be used directly by a different program, three things need to be known: the *values* of the data, the *meaning* of those values and the *structure* of the data. Some standards for an information structure, such as SGML for documents, identify the components of the structure by means of labels. STEP standardizes the *structure* and *meaning* of the information itself. STEP does not replace existing product standards but provides the standardized structures in which data values from those standards can be placed

The core of STEP consists of a set of Parts (41-49) that provide models for the generic description of the characteristics of any product - what it is, its shape, how it was made, etc. These generic descriptions are written in the formal description language EXPRESS, which is both human readable and computer interpretable and is also part of the Standard[6]. The key feature of STEP is that the Parts are connected together to provide a integrated information structure. So the Parts are effectively sections of a large computer programme and cannot be used as individual standards. Compilers and other software tools exist to process the Standard.

The Parts 41 - 49 (4X) of STEP are not intended to be implemented directly. To cope with the details of specific classes of products, STEP uses application protocols (APs)which are also parts of the Standard. An AP can be regarded as a selective filter, specifying those concepts from the Part 4X which are needed for a particular kind of product and giving the generic concepts a specific *meaning* within the chosen scope of the AP (Fig. 1). The APs map the generic data constructs from the Part 4X into the STEP Data File[7] for data exchange or for use in a computer application which recognizes the STEP data structure. If the application is not designed to use STEP data files directly then there is a Standard Data Access Interface or SDAI [8] to enable an interface to be created between the STEP data file and the application.

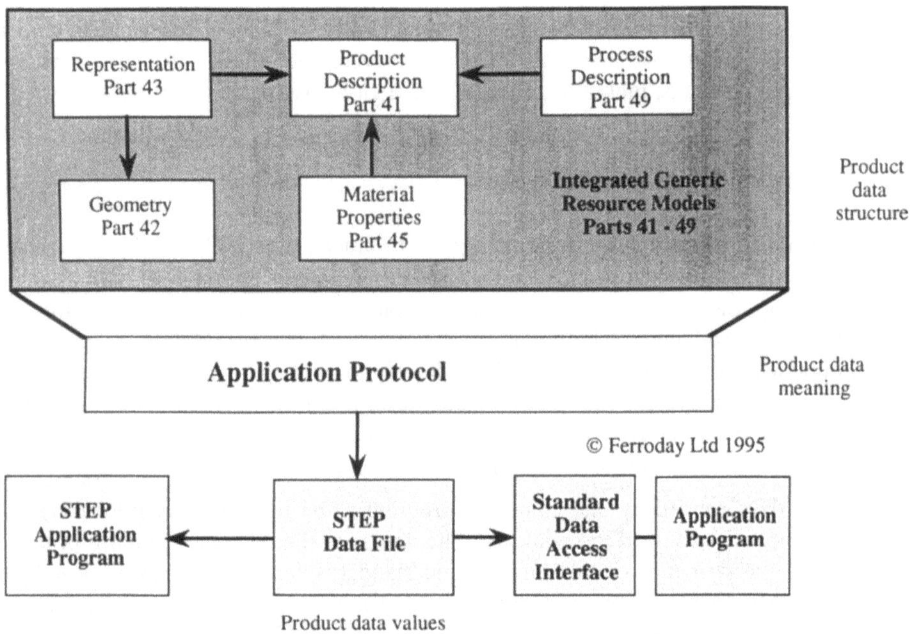

Figure 1. How STEP works

The development of an AP is a task for engineering experts in the chosen field who have to:

- identify the industrial need;
- select the appropriate generic information structures from the central core;
- give them a specific meaning within the chosen scope of the AP.

4 MATERIALS DATA IN STEP

It has been possible to include information on engineering materials into this scheme by recognizing that all engineering materials are in reality products and so could be described by using the resources of STEP. Part 45 of STEP has therefore been developed to provide the generic structures needed to contain the material properties which are characteristic of a manufactured product. This ensures that all the history of the creation of the product can be combined with the material properties which are the consequence of the manufacturing process. Concepts from materials engineering were transformed into STEP by the Materials Task Group in Working Group 3 of the ISO TC184/SC4, responsible for STEP. Part 45 is expected to be issued as a Draft International Standard for international ballot during 1994 [9].

Part 45 of STEP provides generic data structures for :

- the material properties of a product;
- the set of conditions in which the property values are valid;
- the chemical composition of the product;
- the material structure of the product;
- the composition of the components of the structure;
- uncertainty and reliability for numerical values;

The capability to deal with directional structures and directional properties and test conditions is also provided for as a consequence of the integration of Part 45 with the other Part 4X. Future developments will include data structures for properties which can be described in parametric form, such as creep. The existence of Part 45 means that STEP is now capable of more than the exchange of CAD data and can be used to represent data from actual products, including materials products. This development enables the computerized integration of the data in the life-cycle of a product to proceed much further than the design and analysis stage.

The following sections outline the features of the Part 45, describe some of the conceptual problems which the ISO Task Group for Materials have had to resolve and indicate the transformation which this development should bring to material property data.

5 THE MAIN FEATURES OF PART 45

The main source of information on the Part 45 of STEP is the ISO Standard 10303-45 itself and the descriptions which follow are only a simplified summary. In particular it should be realized that no Part of the Standard is independent and several of the important resources needed for a full description of the material characteristics of a product are in other Parts of the Standard. Part 41 [10] is the hub to which the other generic resource models are linked (integrated) (Fig. 1). The central concept of Part 41 is that of a product which has characteristics. One of its characteristics is the shape [11]; another characteristic is the description of the process used to manufacture the product [12]. Part 45 provides the resources for describing those characteristics which relate to the material of which the product is made. The Part 45 is divided into three sections or schemes.

Material Property Definition Schema

This schema provides the means by which :

- A material property can be defined and associated with either a product or a shape of a product;
- A material designation can be linked with: the product, a material property, the composition of a product;
- A material constituent can be associated with a product and the composition can be described;

An important benefit of the integration of the Parts is that the resources provided in Part 42 and Part 43 [13] can be used to describe directional characteristics of a product. This allows for the descriptions of the material structure and for the properties of directional structures manufactured as a composite, for example. The provision for assigning a set of properties to only a part of a shape enables the description of products, such as press-formed parts, that can have different properties in different regions of the product.

The main conceptual problem which arose in designing this section was to retain the concept of a product with a set of characteristic properties, only one part of which are material related properties, and to avoid thinking of the material in the abstract as itself having a set of characteristics which depend on the nature of the product. Hence the term: 'material property' is defined in Part 45 as: 'a product characteristic which depends primarily upon the material or materials comprising the product, the value of which characteristic is determined by measurement or assigned by agreement'. An example of a material property in this context is *electrical resistance* whose value is influenced by the shape and dimensions of the conductor and the resistivity of the material from which it is made. The emphasis is therefore on the properties of the product and not on the properties of the 'material'.

Material Property Representation Schema

This schema provides the resources for representing a numerical value and associating with it the conditions which make that value valid. Several representations of a materials property exist including: numeric values, parametric or fundamental equations, graphical and non-numeric values. The value of a material property may be measured or assigned, by a standards committee for example. If the value is measured then the value will depend on the method of measurement and on the conditions pertaining to the method used. If the value is assigned then the conditions under which that assignment is valid may be specified. In either case, assignment or measurement, the conditions under which the value is valid are expressed as a set of quantitative (numerical) and qualitative (comparative) data which form the 'data environment'. Associations between the data environments are also provided for. The conditions of the method of measurement may be maintained as constant throughout the measurement but some conditions may be varied independently of the others to provide a set of related properties. The data environment may contain qualitative and quantitative data on both types of conditions. Material properties may also be expressed qualitatively. For example, color, although expressible quantitatively as a spectral response to stimulus, is most often described qualitatively by words such as 'red', metallic gray', etc.

Qualified Measure Schema

A physical quantity may have several other aspects other than its value and units. All measurements are subject to uncertainty and values in specifications may be further qualified as maximum or minimum values for example. It is common in specifications of material products to give a range of values over which a material property or a composition is acceptable. The reliability of a numerical value is an indication of the extent to which the value can be used with confidence and would cover such situations as approval for use in design. The subjects of this schema are therefore the resource constructs to allow numerical values to be qualified i.e. further characterized as to their uncertainty and reliability. The treatment of uncertainty which is used in this schema was taken from The ISO Guide to the Expression of Uncertainty in Measurement [14].

6 APPLICATION PROTOCOLS USING PART 45

The basis of STEP is that the generic resources in the Part 4X series should be abstract concepts applicable to any product, without specific meanings being attributed to them. In practice this has turned out to be impossible and all the models are mixtures of abstract and concrete concepts. The problem then arises where in the information *structure* should the *meaning* of a concept be established. The principle of STEP is that the meaning of the concepts for a particular purpose should be established in an Application Protocol and be determined by the scope which fixes the context. It is clear that the level of detail and expression of meaning are closely linked to the scope and deciding on the most effective scope of an AP is very difficult because no guiding rules for determining a scope have been developed yet. The scopes of APs currently under development appear to have been determined mainly by expediency and by practical considerations such as the availability of financial support. It was not possible to develop the Part 45 without also considering how it would be used. At the present time the most common basis of communication of the material properties of products is by the use of national and regional standards. It is assumed therefore that some Application Protocols at least will be derived from existing standards for materials products. Some of the problems which arise with meaning can therefore be revealed by looking at the way in which the constructs in the Part 45 could be implemented in APs.

The material properties of many different products frequently use a common subset of mechanical properties. However there is no universal standard for the names and meanings of these properties and different industrial sectors using different standards interpret these concepts in different ways. Two different APs with apparently different scopes and development routes could therefore contain a set of the same property concepts but in different contexts and, what is critically important, in *different data structures,* as a result of their different origins. The data structures are written in EXPRESS and compiled as code. A supplier of materials data to both of the two different contexts would then have to maintain and supply that data in two distinctly different structures in order to satisfy the same requirement in the different contexts. Also, an engineering analysis program, which would use the common property in different application areas, would not be able to recognize the concepts from the two APs as being the same. This is exactly the situation for which STEP was supposed to provide the solution. It is clear that in order to avoid this problem attention has to be given to the inter-operability of APs and also that different APs which share common concepts with the same meaning in different contexts would have to be integrated, that is compiled, together. There would also have to be a close liaison between the groups developing the different APs. This problem is now recognized and discussions to devise procedures to avoid it have begun. Some of the Application Protocols are listed in Table 1.

Active Projects	
ISO 10303 -	
207	Sheet Metal Die Planning and Design
209	Design Through Analysis of Composite and Metallic Structures
214	Core Data for Automotive Mechanical Design Processes
217	Ship Piping
218	Ship Structures
221	Process Plant Functional Data and Its Schematic Representation
222	Exchange of Product Definition Data from Design Engineering to Manufacturing Engineering for Composite Structures
223	Exchange of Design and Manufacturing Product Information for Cast Parts
Planning Projects	
Data for Engineering Plastics	
Exchange of Design and Manufacturing Product Information for Forged Parts	

Table 1. Some Application Protocols in STEP which use the generic information structures from Part 45

7 ADVANTAGES OF STEP FOR MATERIALS DATA

Anticipating that the problem of interoperability and the use of common meanings in APs will be solved, the advent of STEP will revolutionize the potential for materials data and for materials data systems. The main advantages should be :

- Ready-made data structures which are rich in details;
- The integration of dispersed data systems;
- Flexibility for different purposes;
- Automatic integration with other CAE systems;
- More complete audit trails for the origins of data;
- Increased quality assurance;
- Reduced costs, liability and risk.

STEP only standardizes the data structures and specializes their meaning in APs. There will still have to be a database management system to provide access to the data structure. Some of the structures used in STEP are not easily managed in relational database systems and either object-oriented or hybrid systems are better suited for this task. In order for everyone to benefit from these advantages the main producers of materials data, who are the manufacturers of the materials, will have to recognize that they have an important role to contribute to the development of STEP in their own interest. So far, they are the only industrial sector which is reluctant to participate at the same time as their main customers are strongly committed to this new technology and are coordinating their efforts on a global scale.

CONCLUSION

The development of STEP is a remarkable effort of international collaboration to support the global manufacturing economy with a standard for the open interchange of product data between engineering systems. The process of development requires the cooperation of domain experts from all branches of engineering together with data modeling and computer specialists. A great deal has been achieved and the first fruits of this effort are now starting to appear and to be used in practice. At the same time, some fundamental problems are beginning to be identified which will need to be solved soon. The scale of the effort and the global level of investment in the development of STEP is so large that there is great pressure to put the standard to use as soon as possible, even if the initial versions do contain some flaws and may present some practical difficulties in some situations. What is clear is that the world's manufacturing industry will no longer accept the current lack of standardization of product data and the extra cost burdens that this imposes.

The availability of a set of standard data structures offers the possibility of a revolution in the design and management of large collections of technical data. The weakness of the current situation for materials data is the urgent need to develop Application Protocols, or at least sections of them, which are fully supported by the primary materials producers. Lack of participation in STEP by the manufacturers of materials could lead to a multiplicity of structures and meanings which will make the situation worse for the producers of materials data instead of better.

ACKNOWLEDGMENTS

It is a pleasure to record the cooperation within the ISO Materials Task Group under the Chairmanship of Dr. J. Carpenter Jr., NIST, Gaithersburg, U.S.A. I am grateful to Mr. R.J. Swindells, Editor of Part 45, for guidance on modeling and for helpful discussions.

KEYWORDS

STEP, data, product, data exchange, modeling, international standard, ISO, material, CAD, property.

REFERENCES

[1] ASTM Committee E-49, Computerization of Materials Property Data, ASTM 1916 Race Street, Philadelphia PA U.S.A.

[2] Breuer H., Dean G.D., Progress with International Standards for the Aquisition and Presentation of Data on Plastics for Computerized Databases, in Computerisation and Networking of Materials Databases, Third Volume, ASTM STP 1140, Barry T.I., Reynard K.W. (Eds) American Society for Testing Materials (1992)

[3] Vinard D.R., Pelleing C., Dereims M., Neutral Format for Materials Databanks based on Z99-001 AFNOR Standard, in Computerisation and Networking of Materials Databases, Third Volume, ASTM STP 1140, Barry T.I. and Reynard K.W. (Eds), American Society for Testing Materials (1992)

[4] Breuer H. et al., CAMPUS®- Standardized Presentation of Data on Plastics in: Plastics Gateway to the Future, Proc. of SPE 52nd Annual Conference.

[5] ISO 10303 - 1 Product Data Representation and Exchange - Part 1: Overview and Fundamental Principles.

[6] ISO DIS 10303 - 11 Product Data Representation and Exchange - Part 11: The EXPRESS Reference Manual.

[7] ISO 10303 - 21 Product Data Representation and Exchange - Part 21: Clear Text Encoding of the Exchange Structure.

[8] ISO 10303 - 22 Product Data Representation and Exchange - Part 22: STEP Data Access Interface.

[9] ISO 10303 - 45 Product Data Representation and Exchange - Part 45: Materials Properties.

[10] ISO 10303 - 41 Product Data Representation and Exchange - Part 41: Fundamentals of Product Description and Support.

[11] ISO 10303 - 42 Product Data Representation and Exchange - Part 42: Geometric and Topological Representation.

[12] ISO 10303 - 49 Product Data Representation and Exchange - Part 49: Process Structure and Properties.

[13] ISO 10303 - 43 Product Data Representation and Exchange - Part 43: Representation Structures.

[14] Guide to the Expression of Uncertainty in Measurement, International Standards Organisation, Geneva (1993)

DATA EXCHANGES IN THE BUILDING INDUSTRY

Anne-Françoise CUTTING-DECELLE [1] and Anne-Marie DUBOIS [2]

[1] ESIGEC/LGCH, Campus Scientifique Savoie-Technolac, F73376 Le Bourget du Lac Cedex, France,
E-mail: afcd@univ-savoie.fr
[2] CSTB, BP 209, 06904 Sophia Antipolis Cedex, France,
E-mail: anmarie@cstb.fr

ABSTRACT

The increasing use of computers in the construction Industry has led to a complex situation, where a wide range of software is being developed for different purposes, most of the time without any relation among them.

This situation gives rise to an urgent need for communication and integration, particularly if the profession wants to improve its productivity. On the other hand, a broad effort for standards development is going on, with, among others, EDI, STEP, SGML, PART-LIB. Although these important developments are generally undertaken by large companies, they produce standards which are essential for the construction sector, and industry is now becoming aware of the fact. The purpose of this paper is to present some of these developments, either at a national level, or at a European level, and also to analyze the problem of the data exchanges : where and when they can be found during the overall process of building a structure.

RESUME

L'utilisation croissante des ordinateurs dans l'industrie du bâtiment a conduit à une situation complexe, dans laquelle une vaste gamme de logiciels est développée pour satisfaire différents besoins, la plupart du temps sans aucune relation entre eux : cette situation donne lieu à un besoin urgent de communication et d'intégration, en particulier dans la mesure où la profession veut accroître sa productivité. D'un autre côté, d'importants travaux de normalisation sont en cours, avec, parmi eux : EDI, STEP, SGML, PART-LIB.

Bien que ces travaux soient en général réalisés par de grandes entreprises, les normes qu'ils produisent sont essentielles pour le secteur du bâtiment, et l'industrie en est maintenant consciente. Le but de cette présentation est de décrire quelques uns de ces développements, que ce soit à un niveau national, ou européen, et aussi d'analyser le problème des échanges de données : où les trouve-t-on et à quel moment du processus global de construction d'un bâtiment?

1 INTRODUCTION

Design and construction of buildings is essentially an information process [Syvertsen], particularly in the early phases of design and planning where no physical structure exists, but only related information.

Moreover, the construction process involves, whatever the stage, a lot of people coming from different organizations with various roles. Those persons must work together on the same building, but each of them has his own knowledge, his own language to express this knowledge, his own way of working.

Exchanging information - with the aim of making up ONE construction project - cannot be considered as a trivial task and very often, consequences of misunderstanding may be of great importance to the cost and the quality of the construction.

In this paper, we will try to analyze the communication problem in the building sector, with the specific features of the profession, the type of information to be processed and the corresponding problems.

We will then present the various levels and types of information exchange during the building process.

This paper will end with a presentation of the current situation, the on-going actions and some projects dealing with this matter.

2 THE COMMUNICATION PROBLEM WITHIN THE BUILDING SECTOR

2.1 Specificity of the building sector

The building sector is characterized, in comparison with other manufacturing industries, by an immense diversity of products and processes, and also by a far-reaching separation of design, construction and maintenance. Conventional market practices, and the building industry's overall fragmentation (the companies range from very large and worldwide companies to very small ones) constitute major constraints that inhibit integrated product development.

Transient, or even adversary, relations, rather than "co-makership" prevail among the participants of the building process.

Furthermore, the product to be made is unique, the construction of a new project will need the elaboration of complete new relations between the new, and different, partners of the building project. The team is usually built around a given project and it groups a heterogeneous set of partners.

Compared with other industrial sectors, overall productivity [Laitinen] in the construction sector has not increased. One of the main reasons for this is due to the fact that the quality of the whole construction process has remained at a low value : for example, according to several European studies, the error cost counts for 5 to 10 % of the total building cost, the errors with greatest economic impact occurring in design, project management and on site construction. This is largely due to the low level of industrialization of the building process.

Since the different partners in the building industry are highly inter-related, a simultaneous, or "concurrent" way of working would help to provide an organizational framework for mutual adaptation [10] between the design and construction processes. Such capabilities would allow taking account of the requirements of the construction and maintenance stages from the design stage [CIBAO], [CIB].

Integrated product development is of vital importance to improve the quality of the "building" product, it is also the only way to lower costs and reduce development time.

2.2 Information processing in the construction sector

Whatever the domain, information generally becomes increasingly important as we move towards an information society [ISO TC53]. In the building domain, design and construction are mainly information processing stages : this is particularly true for the design stage, when the building does not yet physically exist - there is only information about it.

Among the **great variety of computer tools**, a disparate set of CAD systems is used, in current building practice, for drawing, geometric modelling and visualization, together with an increasingly wide range of independent, application specific programmes for design analysis and construction, planning and facilities management : isolated application systems, with fragmented data models, and separate file management prevail.

It seems odd to note that nothing has changed with regard to the manual way of working : each area has its own tools - certainly better than many years ago - but without any connection among them !

Integration concerning the application level is just starting. We are at the beginning of CIM systems (Computer Integrated Manufacturing), most of them operational only when devoted to a reduced set of specific tasks.

The building domain also represents an immense amount of information to be managed, when necessary, and where necessary, without forgetting the storing and retrieving devices.

During the construction process [Rio], the project partners exploit existing information, usually coming from other partners, and they can also create new information. These information exchanges give rise to two types of communication problems, one at the platform level (due to the heterogeneity of platforms), the other one at the logical level (need for exchange protocols and semantic models).

Each partner only needs a part of the whole project information; he also needs to adapt it in order to make it understandable by his own tools and he is allowed to enter/modify only a part of it. However, data must be available when needed and the consistency of the whole must be maintained (e.g., information related to maintenance, at the design stage).

For example, the structural engineer mainly needs information related to structural characteristics of the building and to the site where the construction is erected. The architect will provide him with the shape and material data, so that he can carry on his calculations - after an adequate transformation of those data - [Van Nedeveen].

With the development of integration, some requirements are to be taken into account :

- **Necessity for all the required information to be stored in an adequate way**, enabling its use by all the partners involved in the different stages of the construction process, without duplication or loss in spite of the multiplicity of the access points, or the view points of the different partners.

To satisfy this requirement, each stage of the construction process, and all the information flows between them must be identified. Also, since the use of **predefined elements** (such as the components classified in "libraries") is increasing, the system must provide the most proper way of extracting the relevant information. The use of a methodology enabling this functional analysis is indispensable, in order to properly define and situate those different stages.

Furthermore, due to the building life cycle, long term storage (50 to 100 years) has to be ensured, which nowadays is not trivial.

- **Need for technical solution archival and retrieval functions**, because the previous projects constitute the knowledge base of the company.

The last requirement is concerned by the way **systems and tools can be connected :** Not only does it require physical connections between heterogeneous platforms - which is not trivial - but it also requires a high level of semantics.

For example, the transfer of an element called "wall", between two computers, or two applications, assumes that the receptor will understand the name "wall" in the same way as sent by the transmitter, whatever his viewpoint, with :

1) A semantic attached to the name,

2) The same semantic used by the sender.

2.3 Problems arising

Information processing in the construction sector gives rise to a vast number of problems, mainly related to communication, either between computers, or between the different actors involved.

All these points lead to a diminution of the quality of the product to be delivered at the end of the whole construction process, along with an important increase in the cost of the construction to keep the quality at a high level.

Although the information is processed by computers, the essential part of the communication in the construction process relies on human interfaces, [ISO TC59] and the quality of the building can also be associated with the quality of human exchanges during the overall process.

For example, the following situation takes place many times during the construction process : person A (information transmitter), working on a project is about to transfer a message to person B (information receiver), working on the same project. The communication is carried out with the help of "signals", via "media", e.g., drawings, written documents, discussions. Person B must fully understand the contents of the message, i.e. the information, before he is able to react efficiently. He must be able to send back an answer which might be as complex as the request, including results, constraints and modification requests.

3 DIFFERENT LEVELS OF INFORMATION EXCHANGES DURING THE BUILDING PROCESS

In the previous section we pointed out the main levels where the information exchanges take place :

- Definition of the construction process : who is working ? how ? on what ?
- Between the stages defining the whole process,
- With the components to be used and the previous solutions to be retrieved.

3.1 The construction process

Applied to building, as well as to other construction facilities, the construction process is concerned with the complete building lifespan, from the client's first thoughts through demolition, including design, production, operation and maintenance.

For sake of simplicity, this complete lifespan, or lifecycle may roughly be subdivided into (Fig. 1) :

- **creation,**

- **use,**
- **decommissioning.**

Those phases can, in turn, be partitioned into activities and sub-activities according to a taxonomic procedure, e.g. :

- **Inception:** surveying the site, assessing user requirements,
- **Design:** collecting data, synthesizing, drawing, cost estimates,
- **Production:** planning, ordering materials, fixing, finishing,
- **Use:** regulating, cleaning, repairs, maintenance,
- **Decommissioning:** salvaging materials, protecting the public demolishing.

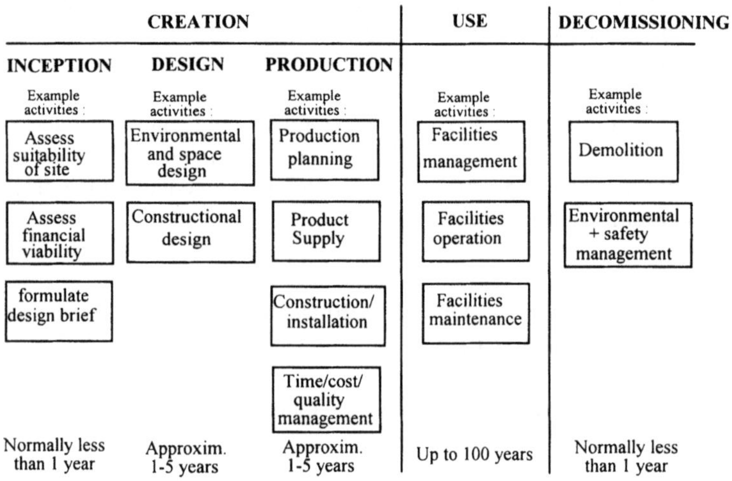

Figure 1. The phases of the construction process [ISO TC59]

The above figure shows that the construction process is long : there are normally at least 50 years between the birth of a project idea, and the remodelling or decommissioning of a typical building. During this time, a massive flow of information involves hundreds of persons from various organizations and with different tasks having to exchange, or store, thousands of facts in relation with the whole construction process. The same characteristics hold for the construction of a new facility or for the alteration, or renovation, of an existing one.

Documents :

One important aspect of the construction process is constituted by the huge number of documents used and generated during this process : traditionally in the form of paper documents, they become more and more "electronic" documents and computer files [Debras, Rezgui]. These documents may be divided into three categories :

- **Project specific documents:** drawings, specifications, bills of quantities, general features, .. Most of them are of major contractual importance,
- **Reference documents:** not specific to a project, but mentioned one way or another in every construction project, such as laws and regulation codes and codes of practice,
- **Technical and commercial documentation** such as supplier catalogs, etc.

Conventional building design computer representations are described in drawings, and related contract specifications, but these do not include all the information needs of the construction process, e.g. the global product information needed for tendering and planning. For example, the available technologies and resources, with associated costs and times, are essential for assessment of constructibility during the entire design process. An integrated approach to building product and process modeling is thus required to effectively support "concurrent", instead of merely sequential development of design and construction. With this approach, the necessary feedback between construction and maintenance, gained from previous experience, requires an integrated approach to support "concurrent" processes.

3.2 Dynamic view of the construction process

The description of the construction process with sub-processes, and tasks, each of them consuming large amounts of economic and physical resources, and transformed into results, leads to the definition of information flow between tasks and identification of actor information requests.

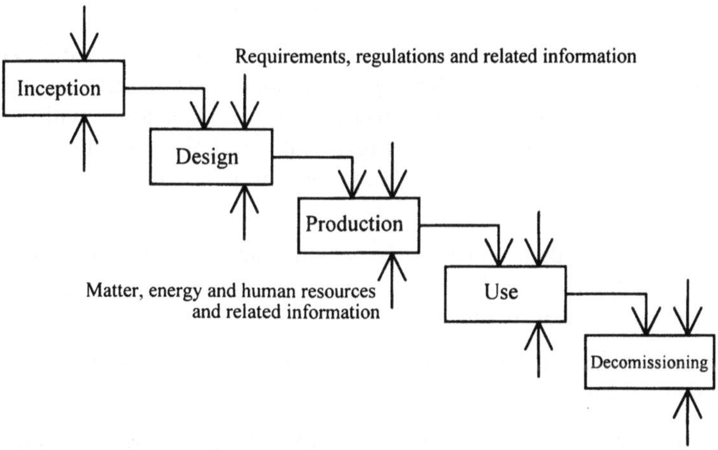

Figure 2. Process model of the construction cycle (rough view)

The diagrams underline both the complexity of the information flow and the strong link of the information flow with the social and economic organization of the construction process, e.g. role and responsibility of the partners.

3.3 Product data model

A product data model is generally defined as "a conceptual description of a product, capable of structuring all the information necessary for the design, manufacturing and use of that product." One model can then be implemented by several mappings in different application programs.

A building product data model should be able to serve as an information framework during the entire lifecycle of a structure and for all the parties involved. No single person holds the entire model in all its detail - it is in fact a distributed model, partially owned and held by each participant in the building act, corresponding to the information flow of the former diagram.

Objective of a product data model :

The **basic objective of a product data model** is to create and maintain a representation of a building being designed, at any stage of product development that is consistent, non redundant and sufficiently complete to be able to contain the information necessary to all the partners involved in the construction, at any time, thus enabling exchanges between computer applications (CAD systems, specific application software), or data integration (CIM systems).

The computer product data model will thus have to provide the capability to represent all the product characteristics over the whole lifecycle, ranging from single parts, or components, assemblages of components to the building considered as a whole, whatever the access viewpoint : structure, HVAC, services, ...

Going from design to construction requires an effective integration, within an overall building information system (fulfilled by the product data model), of the various modelling functions. The current distribution of functions within product developments (CAD, calculations, ...) reflects established divisions among disciplines and vested practices, as well as the limitations of past, and current information technologies.

To date, a number of data models exist which could refer to the product model approach. Most of them are devoted to specific purposes or to a given material: concrete structures or steel structures.

These data models have to be expressed in a formalism such that both humans and computers may read and use them.

Graphical representations such as IDEF1x, NIAM (Nijssen's Information Analysis Method), and formal languages such as EXPRESS are being used in STEP developments.

3.4 Components

Pre-fabricated elements play an important role in the construction industry, and components may represent an important part of standard buildings as, for example, slim slabs, columns, beams, ... for structural elements, or again covering and cladding.

When the parts used in a building seen as a product [ENV 40001-10] are defined independently of the product itself, by a part supplier, the information about these parts, defined in a part library, are independent of any product where these parts may be used.

In this case, information related to these parts must be defined and structured in such a way that it may be exchanged between the various organizations and partners in order to facilitate the generation and the management of information about products containing such parts.

These transfers must be realized independently of any particular system. The nature of the description must make it suitable, not only for the exchange of files containing part libraries, but also as a basis for implementing and sharing databases of parts library data, and archiving.

Problems arising :

The problems arising from the use of components must be dealt with by the building data model as a problem specific to these components.

Indeed, when modeling product data and knowledge [Pierra] in a CIM perspective, two categories of information, data and knowledge appear to be very different :

- **Information referring to company products,**
- **Information referring to the autonomous components used to build up the company products.**

If some points are common to the two categories (the way the information is modeled, for example), others are different, such as :

- **Origin of the information:** components' supplier, instead of company
- **Size and structure of the information:** thousands, for a company, but billions for component libraries: modeling such numbers of parts requires powerful factorizing capabilities
- **Access to information:** direct for company products but by consultation, interpretation, retrieval for components
- **Nature of the information:** component information exists prior to any company product. This information must describe which data will have to be (dynamically) inserted in a product data model to represent each pre-existing component.

Developing an open system architecture for CIM information needs the capability to model, store and exchange both categories of information ; but it also needs to model the relationships between them.

4 CURRENT SITUATION AND ON-GOING ACTIONS

For effective communication between various computer systems, and computer applications, standardized protocols are needed. This standardization may operate at several levels, such as :

- **Company level:** identity of structure between the companies involved in the exchange operation
  - or standardized processes,
  - or standardized messages (documents) formats to be sent, or received,
  - or classification systems (categorization of information according to specific criteria)
- **Product level:** choice of a standardized representation of product data: common product data model
- **Component level:** by means of classification systems

In this section, we will present some actions undertaken at the French, European or international level. These actions can be classified in five categories: message formats, classifications, activity model, product data model, and CAD layers. Where they exist, we present on-going research activities.

4.1 Message format: EDIFACT

Electronic Data Interchange is the electronic transfer of business data [5] between different computers based upon standard definitions of messages. Originally focusing on common exchanges of business operations, such as purchasing, EDI is now moving towards technical areas such as CAD data, product catalogues, ... Organizations have been established to create and carry on these standards at several levels :

- Global: UN / EDIFACT,
- European: EDIBUILD, MD5,
- North-American: CIAG, PDIX, CIMIS.

Current areas of EDI message development within the construction industry include :

- **General communications:** setting up communications, sending lists of message receivers, ...
- **Drawings (CAD):** for CAD data exchange, for sending drawings,

- **Tender documents:** invitation to tender, tender, establish contract, ...
- **Production management:** work done, quantities, ...
- **Trade:** price catalogue, order, invoices, ...
- **Authorities:** permission to proceed, ...

On-going work concerns :

- bill of quantities,
- CAD data.

4.2 Classification systems

Classification systems are aimed at providing detailed categorizations of construction information, according to specific criteria. For example, they categorize types of facilities according to their main purpose (e.g., factories, hospitals, roads, ...), types of spaces according to their use (e.g. office spaces, sleeping or harbor spaces, ...), or types of physical building elements, according to their form or function (e.g., walls, floors, electrical supply elements, ...). These categories make up classification tables which list the code, name and description of each item.

Standardized classification systems provide a mechanism for structuring project information (e.g., documents, databases, models) in a consistent way throughout the industry.

Different organizations are involved in standardization of classification systems, with among them :

- **ISO TC59/SC3/WG2:** aims at the production of an international standard.
- **ICIS:** (International Construction Information Society) international society of organizations dealing with specification systems.
- **EPIC:** European Product Information Co-operation, who have developed a multi-language international product breakdown classification identifying some 2 000 products.
- **CIB:** (Conseil International du Batiment) - International Council for Building Research Studies and Documentation : CI / SfB classification.

4.3 Activity modeling

Process modeling, or activity modeling is the subject of some research. Among them, we can mention :

- **Work engaged at the European level, by the CEN TC310/WG1**, about Enterprise Modeling, whose aim is to carry out a European standard in the domain of "Computer Integrated Manufacturing Enterprise Modeling" [Pr ENV 40003], defined as the joint application of Information Technology and Manufacturing Technology, in order to increase the productivity and responsibility of manufacturing enterprises. It relies upon the CIM-OSA architecture resulting from the ESPRIT I project AMICE.

- **An interesting approach, the BP/RIM project** "Building Process Reference Information Model" [BP/RIM], proposes a common conceptual and terminological framework of understanding of the building process, its "agents" and "information flows" among the participants of the European EDI-standardization EDIFACT Board MD5. This project is aimed at producing a database and a data dictionary including definitions of the phases of the process, the participants, the principles of interaction and information exchanges and the documents to be exchanged among the participants.
- **IRMA** (Information Reference Model for AEC) [Froese] : model resulting from an international electronic conference focusing on high level concepts involved in projects, with a particular focus on the processes involved and their relations to facilities, products, resources, people. It was launched at a CIC workshop and to date involves more than 150 participants. Work is currently in process.

4.4 Product data modeling

Activities of product data modeling probably represent one of the most important actions in the domain of new technologies applied to the construction sector. A number of models are currently under development, as well as for specific or integration purposes. For a given domain, some of them overlap more or less completely, others are incompatible. Others result from synthesis of "elementary" models at a national or international level. Most of the models claim STEP compliance and could be potential input to the standard. Let us introduce some of them, representative of their domain.

French models :

The French Ministry in charge of Construction is funding a number of projects dealing with building data modeling in the framework of Plan Construction and Architecture (PCA):

- **MOB:** object oriented building model, The main specificity of the project is that it is run by representatives of contractors, e.g. Bouygues, engineering, e.g. OTH and architects. It aims at the definition of an interprofessional reference model which will be built and validated on an example of building, the "standard building". The models are expressed in NIAM and EXPRESS.
- **SUC:** standard aimed at exchanges between CAD tools, SUC (Standard Unifié de Communication) is a second example of a project run by

professionals (CBC, DUMEZ, SAE). It uses the layer concept as a support for a data structuration enabling the definition of characteristics of the building elements in the design phases together with the corresponding graphical data.

- **BOX:** exchange standard relying on a relational database. BOX is built up on SUC results. It is interesting to notice that this project involves six CAD software developers in a cooperative effort to enrich the capabilities of the SUC standard. The current state of the project proposes a way to exchange not only the data but the models used to describe those data.

Synthesis models (French level) :

- **GSD Project** (Groupe de Structuration des Données): funded by the French Ministry of Construction, is defined as a synthesized effort of the French research teams involved in the domain of building modeling. Different models were analyzed, including architectural design, integrated design, energy, building project management, energy simulation. The GSD model focuses on the design stage with multiple points of view.
- **SIGMA:** is a new project launched by the French Ministry of Construction, whose objective is to formulate a Building Data Reference Model for the French Construction Industry. This project will be built on the results of the GSD project but will also integrate a wider range of recent modeling developments. It is proposed that this project adhere to the STEP methodology and its results be delivered to the Building AEC working group (ISO TC184/SC4).

European projects *(not exhaustive list)* :

- **ATLAS (ESPRIT):** whose objective is to demonstrate and evaluate how large scale engineering projects (typically buildings and process plants) can benefit from the application of advanced and integrated Information Technologies.
- **CIMSTEEL (EUREKA):** CIM system aimed at the steel work industry, to harmonize the data exchanges between the different activities of the whole process of building a steel work structure.
- **COMBI (ESPRIT):** whose aim is to develop a prototype of a computer-integrated environment for cooperative design and concurrent engineering in the building industry - main idea: integrated by communication.
- **COMBINE (JOULE):** Integrated Data Model, aimed at the communication and data exchange between design tools - long term objective : to build on an Intelligent Integrated Building Design System.
- **PISA (ESPRIT):** whose goals are to develop and demonstrate a methodology for information modeling and product/process modeling,

interoperability of applications, model integration, implementation and validation.

Ongoing standardization activities:

- **CAD/LIB:** about components library (not specifically building) started as an ESPRIT project, and leading to the CEN Pr ENV 40004 standard. This work is now carried out at the international (ISO) level, and will lead to the ISO 13584 P/LIB standard.
- **B-C APPP** (Building and Construction Application Protocol Planning Project) launched in October 1993 within the framework of the TC184/SC4/WG3/Team12 STEP-AEC, it aims at defining the scope and developing a framework for the development of a consistent suite of STEP Application Protocols in the building domain. This project is intended to consider the building in all its diversity, with all its "actors" and during all the phases of the building process. A main objective will be to facilitate co-operation between current [collaborative research, national research, industrial and software vendors] interests and establish how the various initiatives can and will contribute to the application of STEP to the Building Industry. Numerous teams, all over the world, are involved in its development.
- **PDBuild** is a new work item defined within the CEN TC310/WG2 working group. Its scope is to contribute to the STEP product data modeling developments and dissemination, in the European Construction Industry. It may also be considered as a shift between national and international levels, a gateway enabling the development and test of common views.

4.5 CAD layers

In contrast with the advanced conceptual modeling methodologies followed in efforts such as STEP, [Froese], a typical current practice for attaching semantic meaning and classification breakdowns to project information is to simply place different graphical CAD objects on different layers of CAD drawings. The naming schemes for the layers describe the various dimensions of the information categories they contain, e.g., building elements, life cycle phases, agents, ...CAD layering is used by specific companies, CAD user groups or sometimes nationally. An international standard for using layers in CAD for construction is currently in progress (ISO TC10).

CONCLUSION

As with other industrial domains, the Building sector more and more depends on Information Technologies. This need for an ever increasing number of exchanges, increase in message numbers and complexity, in the quality and rapidity of transfer cannot be satisfied without corresponding technologies.

The erosion of cultural and political differences between countries coincides with astonishing advances in the power and sophistication of computer systems and applications. In the future, Information Technology will rely upon a high level international integration via protocols such as EDI and STEP.

Product Data Technology [Cutting] offers exciting possibilities for the future but also presents a major challenge : if it is clear that the operational benefits will probably be substantial and various, it is nevertheless also clear that the creation of universal models will require investments of substantial resources. It is thus vital for the modeling work itself to be founded on sound basic precepts, in order to make sure that the direction opted for either results in some benefits or, at least, does not lead to a situation worse than before or to a deadlock.

KEYWORDS

STEP, data, product, data exchange; modeling, component library, international standard, ISO, life cycle, CAD, construction, building, construction process.

REFERENCES

[Augenbroe] The COMBINE project. Proceedings of the Combine Seminar, Stuttgart (1992)

[BP/RIM] Hauch P., EDIBUILD BP/RIM, BPS-centret

[CIB] Computer Integrated Building, Strategie Final Report, (Dec 1991), A.Dupagne Ltd. - Esprit II Exploratory Action N°5604

[CIBAO] Dubois A.M., Prototyping approach of multi-actors,computer aided design for buildings in the frame of the CIBAO project. Proc. of MICAD 90, Paris, (12-16 Feb. 1990)

[Cutting] Cutting-Decelle A.-F., Dubois A.-M., Product Data handled by STEP: Considerations and Proposals, Proceedings of European Product Data Technology Days, MICAD, pp 325-337 (1994)

[Debras] Debras Ph., Rezgui Y., Software systems for the integration of documentary engineering within a construction project. CIB W78, Workshop on Integrated Construction, Helsinki, Aug.(1994)

[Froese] Froese T., Summary of Stockholm Workshop, email , (April 1994)

[ISO TC184/SC4] ISO TC184/SC4, ISO 10303 standard (1994)

[ISO TC59] ISO TC59 / SC13 / WG2, Classification of Information in the Construction Industry, ISO Technical Report (Type 2), Dec.(1993)

[Laitinen] Laitinen J., Cocon Project (1993)

[Pierra] Pierra G., Modelling Classes of Pre-existing Components in a CIM Perspective : the ISO 13584/ENV 40004 Approach, Proceeding of European Product Data Technology Days, in Revue internationale de CFAO et d'infographie, pp 435-453, March (1994)

[Pols] Pols A. A. J., Towards Integrated Support for Computer Integrated Building, pp 41-50, Elsevier (1992)

[Pr ENV40001-10] CEN TC310/WG3, CAD Parts Library (CAD-LIB), conceptual model of parts library (1993)

[Pr ENV40003] CEN TC310/WG1, Computer Integrated Manufacturing - Systems Architecture - Framework for Enterprise Modelling, DS/INF 61, Kóbenhavn (1990)

[Rio] Rio O., Recuero A., Alvarez M., Suarez A., Towards the Integration of the Design Process.

[Syvertsen] Syvertsen T. G., Object Oriented Product Modelling, ARECDAO, Barcelone (1991)

[Van Nederv] Van Nedeveen S., View Integration in Building Design, Management of Information Technology for Construction, pp 209-222, Singapore (1993)

NEUTRAL PRODUCT MODELS FOR SHIP HULL GEOMETRY AND THEIR USE IN DATA EXCHANGE

Horst NOWACKI, Uwe LANGBECKER and Christof REHLING

Technische Universität Berlin, Institut für Schiffs- und Meerestechnik, Salzufer 17 - 19, 10587 Berlin, Germany

ABSTRACT

This paper reports recent developments in the shipbuilding industry toward a neutral description of the geometry of ship hulls on the basis of the product modeling standard ISO 10303 (STEP). In ongoing German research projects first experiences are gained with these models and their use in prototype implementations for data exchange between heterogeneous shipbuilding CAD systems. Simultaneously international standardization advances toward the ISO Application Protocol for Ship Molded Forms.

RESUME

Une description neutre de la géométrie des carénages de bateau sur la base du standard de modélisation d'un produit ISO 10303(STEP) est présentée dans cet article. Les premières expériences des projets allemands de recherches en cours sont suivis avec ces modèles et leur utilisation dans l'implantation de prototype d'échanges de données entre divers systèmes CAD de construction navale. De façon simultanée, la standardisation internationale progresse vers le protocole ISO application pour les formes moulées en construction navale.

1 INTRODUCTION

The exchange of digital product data in standardized form between business partners in the shipbuilding industry and in support of the process chain within the enterprise offers important economic and organizational advantages. The development, production and operation of a ship involves a multitude of partners and decision-makers usually equipped with many different CAX systems and product data representations. The use of neutral product models based on the forthcoming international standard ISO 10303 (Standard for the Exchange of Product Definition Data, STEP [1] promises to provide a stable solution for the communication requirements between heterogeneous systems in the shipbuilding industry.

The geometry of the ship hull, including the ship's exterior hull shape, its principal decks and its main appendages, but not its interior subdivision and compartmentation, is a key element of information in the ship's product model as a reference for numerous applications. The idealized reference geometry of the hull, which consists of the surfaces and curves to which the steel parts of the vessel are to be attached, but which does not account for the thickness of any part, is called the molded geometry of the ship. This is in accordance with traditional steel shipbuilding practice where this geometry was laid out in a workshop called mold loft. This is why the product models being developed for this purpose are called "Ship Molded Forms".

This article gives an overview of three stages of this development leading up to a future ISO Application Protocol (AP 216). A first prototype model for Ship Molded Forms and a test implementation were completed in the German research project K2 in the period of 1991-92 [2]. Currently a revised and somewhat extended model is being tested in the ITiS project [3] in Germany. Simultaneously an ISO AP for Ship Molded Forms is under development, though still at an early stage, under the guidance of the ShipSTEP project and the classification society Lloyd's Register. This initiative at ISO level aims at a definitive and comprehensive solution for this product model, which must be fully harmonized with other shipbuilding APs in progress.

The initial pilot developments in K2 and ITiS met with the difficulty that no complete and consistent STEP based methodology existed for the development of a set of APs belonging to the same modeling domain, hence strongly depending on each other. This posed a major integration problem for the family of shipbuilding APs similar to the situation encountered by the automotive industry within their large single AP 214 (Core Data for Automotive Mechanical Design Processes). At the same time it is an important requirement to support various usage scenarios for shipbuilding applications that require the formation of functional subsets within one or several APs. The modeling tasks thus require special consideration for the methods of integration and modular subdivision.

This paper summarizes the developments toward a modeling methodology for Ship Molded Forms as one in a family of shipbuilding APs designed to support several usage scenarios. It describes the present product models for Ship Molded Forms and future requirements in an ISO context. Finally it discusses current implementation experience.

2 Modeling Methodology

2.1 AP guidelines

The ISO document "AP Guidelines" (full title in [4]) defines the concepts and components of STEP Application Protocols and explains the process of AP development.

An Application Protocol describes the scope and structure of an information model for a specific product model application context.

This AP includes inter alia :

an Application Reference Model (ARM), which describes the product model in terms of application-specific concepts and terminology,

an Application Interpreted Model (AIM) which uses integrated resource constructs to express the information contents of the product model by specializing and constraining the integrated resources to meet the AP requirements.

A mapping table serves to document the correspondence between AIM constructs and ARM constructs.

However the AP Guidelines and other STEP documents do not provide any guidance regarding the internal structure of an ARM, the integration of functional subsets from one or several APs into a new application for a certain usage scenario, nor regarding the modular subdivision of large APs.

AP Guidelines do offer the notion of Units of Functionality (UoF), i.e., collections of coherent application objects at ARM level, which belong together from an application viewpoint. But UoFs are subsets of the ARM, often of major scope, which serve as the only available aggregates for defining the scope of new applications. This mechanism may not be flexible and selective enough to define new usage scenarios.

This is one reason why many developers of large APs or AP families with numerous usage scenarios have found it necessary to develop their own methodology for AP integration and modularization.

2.2 Shipbuilding reference model

The ITiS Project has developed a reference architecture for shipbuilding product models [5]. This reference model takes a macroscopic view of the product model structure by subdividing it into disjoint sets of application-specific modeling subdomains, called partial models, at the ARM level (Fig. 1).

This global view of the entire modeling domain secures the completeness of the model scope and assists in harmonizing the results of concurrent AP developments.

The reference architecture distinguishes four main classes of partial models:

A. Global Characteristics Class: Information about the ship in total (like principal dimensions, hydrodynamics ...)

B. Physical Characteristics Class: Information about physical parts of the ship (like shape, material ...)

C. Functional Characteristics Class: Information on the functional role of shipboard components in their system context (like pipe flow coefficients)

D. Supporting Class: Information by other classes (like arrangements, molded forms etc.)

Partial models may encompass one or several EXPRESS schemas. The schemas should be as self-contained as possible, but will make occasional use of entities from other schemas. Application level schemas (ARM) in particular consist of entities that are mappable onto entities from resource schemas (AIM).

2.3 Shipbuilding application protocols

The Committee ISO TC184/SC4 has established the following shipbuilding APs as work items:

AP 215: Ship Arrangements

AP 216: Ship Molded Forms

AP 217: Ship Piping

AP 218: Ship Structures

AP 226: Ship Mechanical Systems

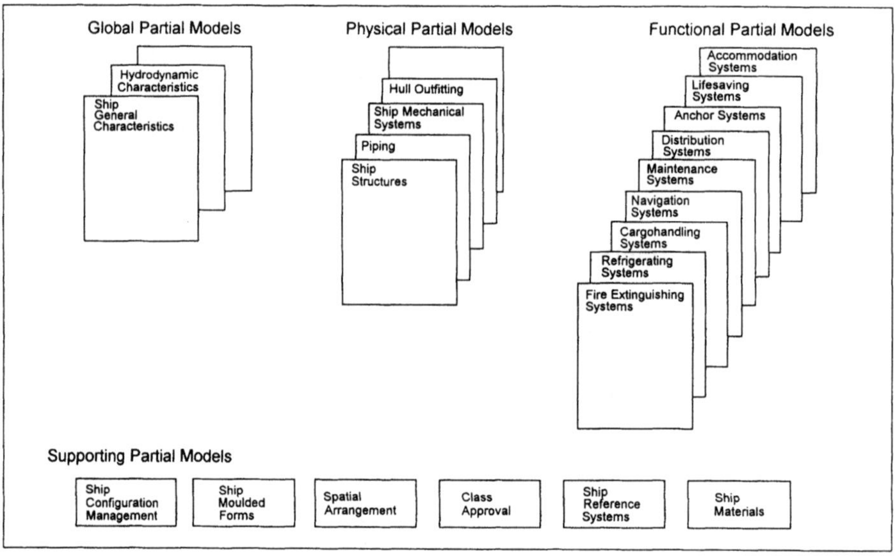

Figure 1. Partial Model Structure for the Shipbuilding Product Model

It should be noted that these AP domains are strongly interrelated so that the APs must refer to each other and share common resources. For example, pipes are attached to structures and both refer to molded geometry. The reference architecture with its disjoint partial models may assist in identifying and removing redundancies in the structure of the APs.

2.4 MARITIME building block methodology

The ESPRIT III Project No. 6041 MARITIME has developed a modeling methodology for STEP APs that supports strong modularity and local information shielding. It is based on the concept of a "Building Block (BB)", i.e., an information aggregate organized as an application-specific schema, which is sandwiched between an import and an export schema so that access to the building block is limited to certain open paths.

The BB approach promotes strong modularity at the ARM level and assists in avoiding conflicts in concurrent engineering developments. However it does not provide a top-down perspective and an integration policy across a set of APs. An integration methodology should seek compromises between the principles of the reference architecture and the requirements of the BB approach. A suitable set of building blocks should be tailored to fit into the partial models and AP architectures without redundant information for any application.

Figure 2 shows an attempt to associate several current building blocks of AP 218 (Ship Structures) with partial models in the ITiS reference architecture. It demonstrates that a unique association without conflicts and redundancies appears feasible.

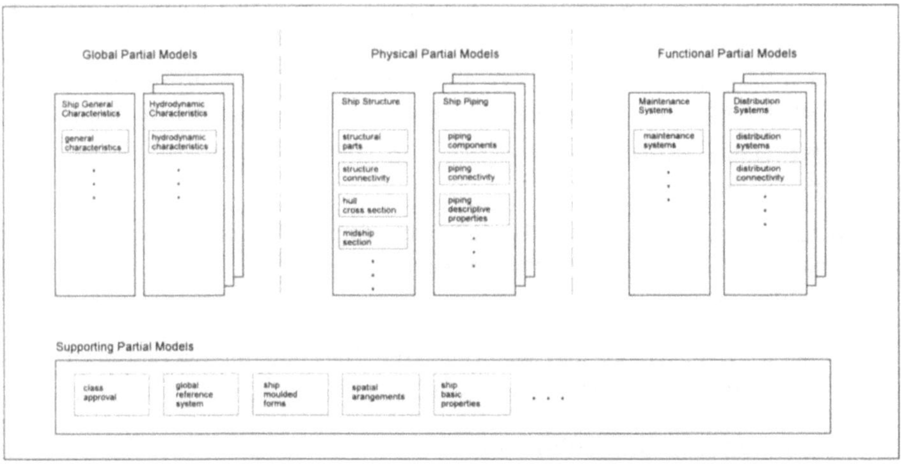

Figure 2. Assigning Building Blocks to Partial Models (Examples)

2.5 Usage scenarios

The scope of a major Application Protocol often attempts to cover the contents of a product modeling domain rather systematically and completely. In a specific data exchange application it frequently happens that only a limited functional subset of the full AP is intended to be exchanged to support several different usage scenarios with specific application contexts. Therefore the ability to select functional subsets from the ARM of an AP as flexibly as possible in order to support diverse usage scenarios is an important modeling requirement.

For the AP Ship Molded Forms it was tentatively concluded on the basis of a system survey that the following levels of geometric functionality will be representative of the main groups of usage scenarios (or user constituencies):

- Point based models (offset tables)
- Wire-frame models (curve mesh topology)
- Surface models (surface topology)

The higher quality models fully encompass the low functionality ones. This suggests forming three usage scenario classes for the geometric functionality in Ship Molded Forms as shown in the following table.

Usage Scenario Class	**Geometric Functionality**		
	Offset Table	**Wire-frame**	**Surfaces**
G 1	✓		
G 2	✓	✓	
G 3	✓	✓	✓

In addition the scenarios may or may not require the addition of hydrostatic data, which may be associated with geometry. This can be denoted by the terminology:

HS1: with hydrostatics
HS2: without hydrostatics

In conclusion, a usage scenario class may be composed of combinations of these functional subsets and may be denoted by, e.g.:

- G3 - HS1

Clearly not all combinations may be meaningful. Thus only the relevant main usage scenarios should be standardized.

2.6 Requirements and solutions

In order to supplement the modeling methodology based on STEP AP Guidelines and in order to promote the integration and modularization of product models at the ARM level the following additional <u>requirements</u> shall be met :

- The overall modeling domain of a major AP or a set of APs shall be clearly subdivided into subdomains or partial models which are disjoint subsets of the full domain. A reference architecture is a useful means for documenting this decomposition. APs shall be associated with one or several partial models.

- Strong modularity of the contents of an AP shall be promoted. A subdivision into several schemes as in the MARITIME building block approach shall be encouraged.
- The support of several usage scenarios by one AP or by a set of APs necessitates the flexible formation of functional subsets within APs. This is facilitated by product models with strong modularity.
- The leaf elements in the ARM information structure, i.e., the smallest information aggregates which serve as ARM primitive elements, shall be of such type and structure that they can be readily mappable onto specializations of resource level (AIM) constructs. This promotes simple mappings.

A solution to these requirements can be achieved if the modeling methodology encompasses the following capabilities and features shown in the AP architecture of Figure 3 :

- ARM/AIM separation by clear distinction between application-specific (ARM) and resource level (AIM) entities and schemas.
- At the ARM level larger aggregations of information (Units of Functionality) are identified as schemas or entity clusters. These larger granularity aggregations are composed of very small granularity primitive level aggregations (called Primitive Units of Functionality or PUoF). The PUoFs are still application-specific, e.g. ship_waterline (WL). They occur as leaves in the ARM information structure.
- A Usage Scenario Class is defined by collecting one or several UoFs. Thereby the required set of PUoFs is identified, too.
- The root level entities or schemas of the AIM are the access points to the resources. PUoFs are mappable onto resource root constructs, preferably by simple mappings (1:1 or 1:n).
- At the AIM level Conformance Classes may be formed as aggregations of resource constructs. The functional scope of a Conformance Class may be made equivalent to the scope of a specific Usage Scenario Class, but this is not necessary. The scope of Usage Scenarios is governed by application requirements whereas Conformance Classes are destined to reflect common test requirements and procedures. Whether each usage scenario needs to have its own Conformance Class, seems to be an open ISO policy issue, on which there are pros and cons.
- If several Usage Scenarios share some PUoFs, it is obvious that they will share some of the resources. This will allow the use of Application Interpreted Constructs (AIC). In any case, the degree to which resources are shared determines the degree of interoperability between applications.

3 Product Models

3.1 Overview

The shipbuilding industry has been an early participant in the development of STEP product models. Many of these models have concentrated on physical and functional aspects of the ship, but all included at least a nucleus of ship hull geometry, which is always needed as a reference for shipboard systems.

The U.S. project NIDDESC, which has been active since around 1988, was the first shipbuilding project that made contributions to STEP, mainly on ship structural systems and ship distribution systems [6], [7].

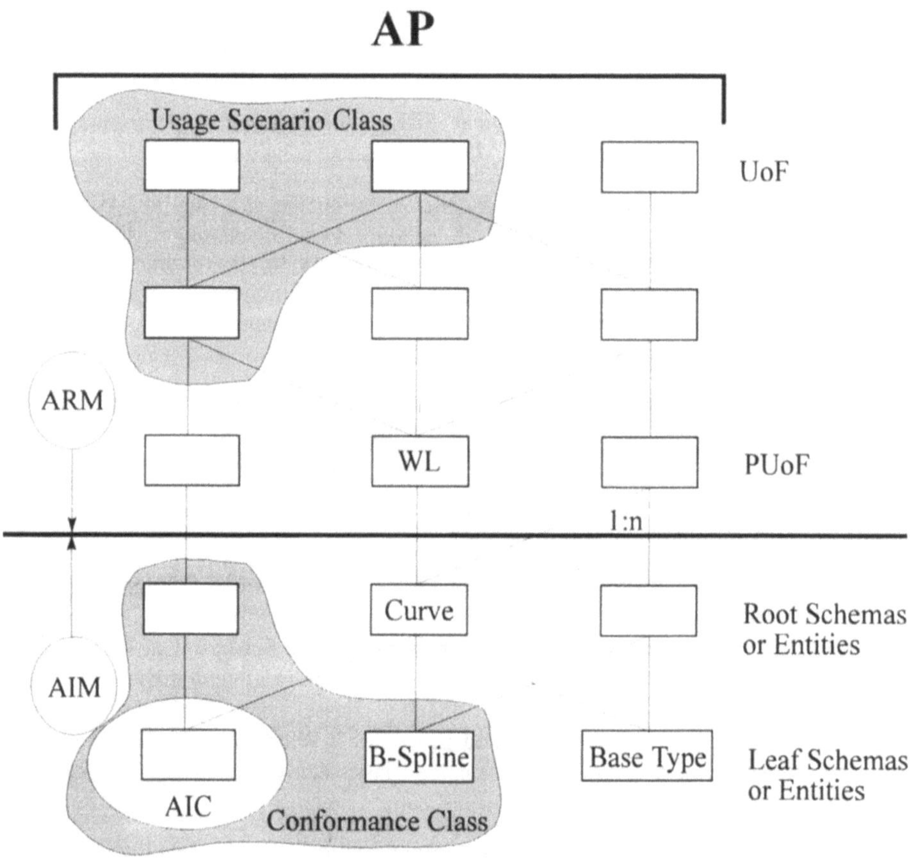

Figure 3. Proposed Application Protocol Architecture

Abbreviations:

AIC: Application Interpreted Construct

UoF: Unit of Functionality

PUoF: Primitive Unit of Functionality

WL: Waterline

The ESPRIT II project NEUTRABAS [8], carried out between 1989 and 1992, was the first European project on this subject. It dealt primarily with product models for ship structural and ship outfitting systems (like NIDDESC), but also included partial models for ship arrangements and ship hull geometry. NEUTRABAS also developed a first Ship Application Reference Model as a tool for integration [8].

In the ESPRIT III project MARITIME [9] the main focus is again on product models and STEP APs for ship structures and ship arrangements. Ship hull geometry is adapted to these purposes from STEP resource parts where it is needed. MARITIME uses ship geometry, but does not aspire to present it in self-contained form in the manner of an AP.

A primary emphasis on Ship Molded Forms product models was placed in the German research projects K2 [2] and ITiS [3]. The results will be reported in more detail below. Recently a work item was established by ISO for an Application Protocol 216 "Ship Molded Forms". This ISO activity is led by Lloyd's Register, U.K. in the context of its ShipSTEP project. Since the STEP resource parts on geometry/topology, mainly Part 42 of ISO 10303, have become mature and stable as International Standards, it will now be possible to develop a definitive form for the AP Ship Molded Forms as part of a shipbuilding AP set. This work is still in progress and the current experiences from K2, ITiS and other projects will serve as inputs to these developments.

3.2 The K2 model

The Ship Molded Forms developments in project K2 have resulted in a prototype product model and a test implementation to support the data exchange of hull geometry between heterogeneous shipbuilding CAD systems for hull form design and analysis. The functionality of this model, which is described in more detail in [2], can be briefly characterized as follows:

- Geometric elements:

  Point sets, point vector sets, curve sets, curve meshes, surfaces.

- Topological elements:

  Offset tables, curve segments, knots, meshes, surface patches, curves on surface, trimmed surface.

- Form:

Single schema ARM with AIM mappings.

- Organization structure:

  Hierarchical level structure decomposed into hull, hull zone, hull zone representation (point list, curve set, curve mesh, surface) with primitive elements that can be mapped onto Part 42 resources.

- Usage scenarios:

  Essentially one scenario: Data exchange between ship geometry supporting CAD systems.

Figure 4 gives an overview of the basic structure of the K2 model. This model meets the needs of a simple data exchange between shipbuilding CAD systems as was demonstrated by test implementations [8]. But the K2 model does not have the deeper, more complex structure required of an ISO AP.

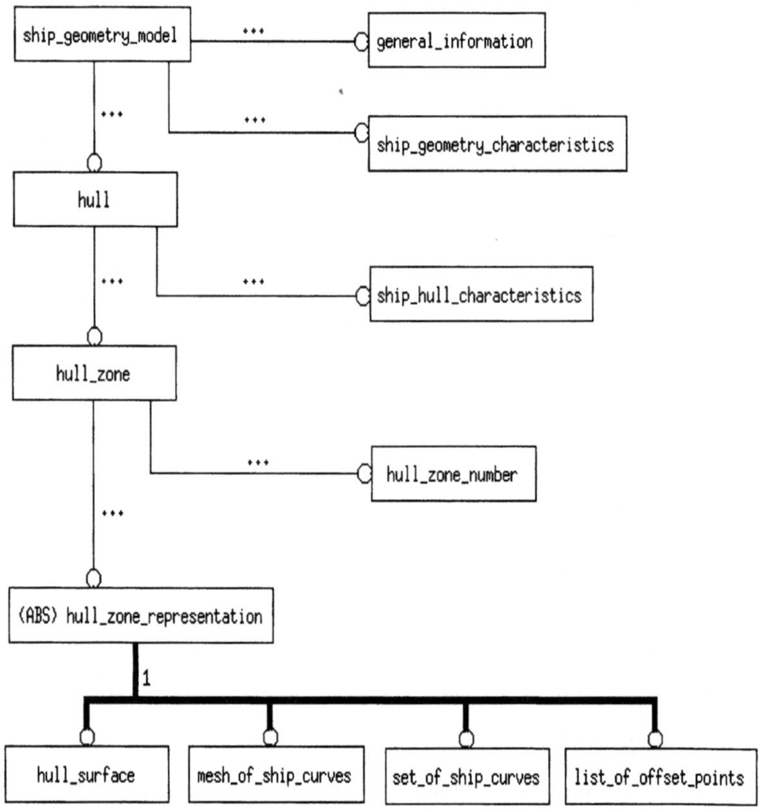

Figure 4. Basic Structure of K2 Model (Simplified)

3.3 The ITiS model

The ITiS sub-project ITiSA is developing another Ship Molded Forms product model that will serve as a hull geometry data exchange platform, but also as a building block in the mosaic of other ITiS partial product models. This new model is a refinement and extension of the K2 model with a more complex information structure (Fig. 5).

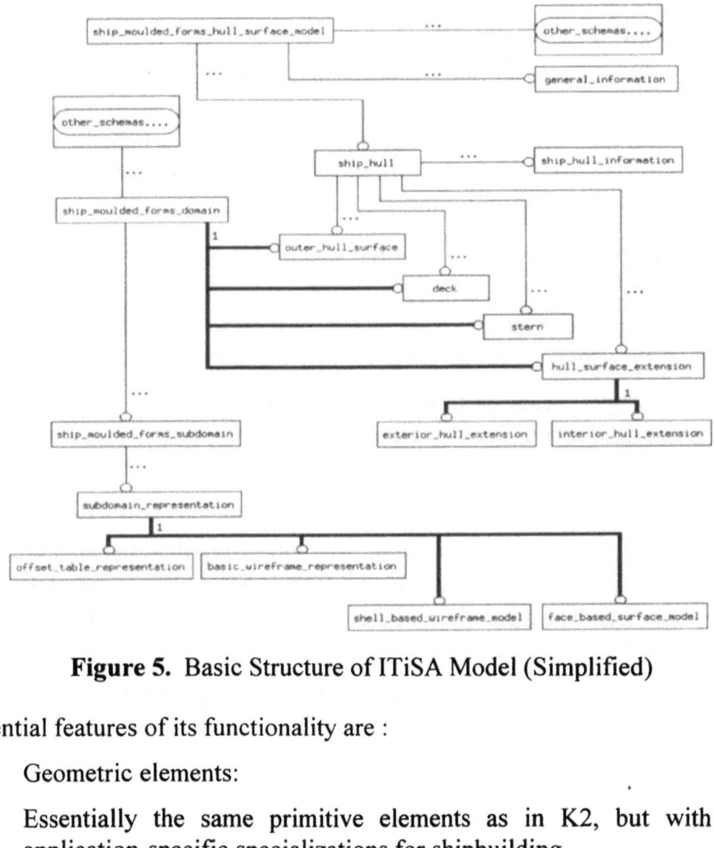

Figure 5. Basic Structure of ITiSA Model (Simplified)

The essential features of its functionality are :

- Geometric elements:

  Essentially the same primitive elements as in K2, but with stronger application-specific specializations for shipbuilding.

- Topological elements:

  Essentially the same scope as in the K2 model.

- Form:

  One main schema for Ship Molded Forms augmented by 11 other schemes for administrative and other general purposes (coordinate systems, symmetry properties, measures, general characteristics etc.). This reorganization is a refinement of the K2 model made in the interest of greater flexibility and modularity. Many new schemes resemble

MARITIME Building Blocks and will in the future be replaced by corresponding ISO AP schemes.

- Usage scenarios:

  The current ITiSA model is capable of supporting up to three usage scenarios at the levels of offset tables, wire-frames and surfaces.

3.4 The ISO AP 216 model

This model is still in the early stage of its development within the Ship STEP project. Currently the requirements are under discussion and an Application Activity Model (AAM) has been proposed by Ship STEP for this AP. Certain inputs can be made to the requirements on the basis of the experiences from K2 and ITiSA. In brief summary the main requirements for this future AP include:

- The geometric functionality shall encompass at least the same scope as in K2 and ITiSA.
- The topological structure for hull form representation shall also include at least the same options as in the former two models.
- The Usage Scenarios to be supported by AP 216 shall comprise at least those three main classes identified earlier in section 2.5. This should be reflected in the internal structure of the AP.
- The leaves of the ARM should be readily mappable onto AIM resources.
- Selective use of entities from the Ship Molded Forms AP shall be convenient so that other APs can make effective use of ship geometry constructs. The strongly modular structure outlined in section 2.6 may be useful for this purpose, too.

Current discussions about AP 216 are addressing these requirements.

4. IMPLEMENTATION EXPERIENCE

4.1 Software architecture

During the development of data exchange software implementations based on STEP in which we were involved in various projects including NEUTRABAS, K2, ITiS and MARITIME, we experienced rapid developments of software tools and went through several generations of implementation environments. In the earlier projects NEUTRABAS and K2 the STEP standard was still unstable and most tools went through dynamic changes.

In these projects our partners had to develop their own implementation tools because no suitable ones were available outside. This was done by Digital Equipment in

NEUTRABAS [8] and by Ingenieurzentrum Schiffbau, Rostock and ZGDV, Rostock for K2 [2]. The experience in both cases was quite successful, though much effort went into these developments, and the results were demonstrated in prototype data exchange environments. The experiences were later used by these partners to develop more definitive software tools.

The more recent projects ITiS and MARITIME have commercial software tools available and are making use of such tool-kits. The functionalities of these implementation tools can be judged against a reference architecture (Fig. 6) which attempts to concentrate on the common denominator of many implementation environments.

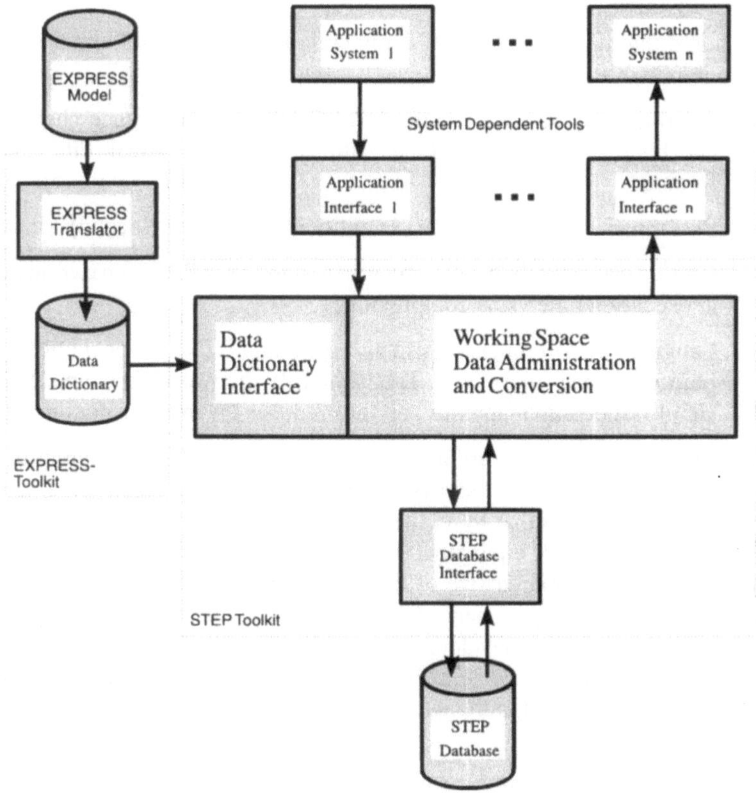

Figure 6. Reference Architecture for STEP Based Data Exchange (Late Binding)

An EXPRESS tool-kit is required to translate the EXPRESS product model, to verify its syntax, and for late binding tools to create a data dictionary metamodel which will serve as a reference in working space. Secondly a STEP tool-kit is needed to administer the data exchange and conversion process in working space and to organize the communication with the STEP repository or database. Finally system-dependent conversion tools need to be developed for each system to read and write between application system and STEP working form. To develop these is the main system-dependent implementation task for each application system.

4.2 Tool-kit experience

Several tool-kits are available to the shipbuilding modeling teams involved in the current AP and prototyping developments. They include ITI, NIST, ProSTEP and STEP tools. Although most tool-kits are functionally stable, further developments are often still in progress so that comparisons remain tentative. Despite that certain evaluations are possible.

We have found that in comparisons of commercially available tool-kits the following criteria are important:

- Conformance with the standard (at the levels of reference checking, type checking and value checking with regard to Physical File and SDAI formats).
- Performance (measured, e.g., in entity operations per second)
- Late vs. early binding (i.e., the processing of EXPRESS based instructions at execution time vs. at compile time)
- Language environment (usually procedural like C or FORTRAN vs. object-oriented like C++. The similarity between C++ and EXPRESS simplifies certain mappings and thus reduces the implementation effort).

In the current developments toward a Ship Molded Forms data exchange in ITiSA the ProSTEP tool-kit was chosen. It shows excellent conformance with STEP. ProSTEP is a late binding tool-kit so that by virtue of its more intensive processing load at run time it does not reach fully equivalent performance compared to early binding tool-kits. But in view of the limited size of the product model instantiations which occur here this is no serious impediment. Moreover model updates and changes, which occur frequently during our development, can be flexibly accommodated by a late binding tool-kit because it is not schema-dependent. The embedding of the ProSTEP tool-kit in C was accepted although a C++ solution would have offered some development shortcuts.

The implementations are still underway and further experience will be gained.

CONCLUSION

This paper gives an intermediate report about the developments in progress toward a shipbuilding data exchange for the product model dealing with ship hull geometry.

The product models for this application domain are undergoing revision and reorganization in order to bring them from a single-schema, single usage scenario level (as in K2) gradually to a multiple-schema, multiple usage scenario quality (via ITiS), which will be required when the product model is embedded in the complex mosaic of a coherent set of ISO shipbuilding APs.

The paper describes the principles of an AP modeling methodology as a top-down strategy toward integration and modularization. It documents the modeling requirements and certain solutions resulting from this. It presents the current results of these developments and reports some of the major implementation experiences gained so far.

KEYWORDS

Product modeling, product data technology, ship form data model, STEP software tool-kits, neutral interfaces, data exchange, shipbuilding application protocols.

REFERENCES

[1] Owen J., STEP, An Introduction, Information Geometers Ltd., Winchester (1993)

[2] Nowacki H., Stolte F., Klingner J., Vatterrott G., A Neutral Product Model for Ship Hull Geometry and its Use in Data Exchange, European Product Data Technology Days 94, Revue internationale de CFAO et d'infographie, vol. 9,n° 3, Editions Hermès, Paris (1994)

[3] Stolte F., Nowacki H., Information Model for the Exterior Hull Geometry, Task 1.1.2: Intermediate Report, in German, TU Berlin, June (1994)

[4] Palmer M., Gilbert M., Guidelines for the development and approval of STEP application protocols, version 1.2, ISO TC 184/SC4, PMAG N506, May (1994)

[5] Langbecker U., Nowacki H., Bühr W., Kress H., A Reference Architecture for Shipbuilding Product Models, Intl. Conf. on Comp. Appl. in Shipbuilding, Bremen, Sept.(1994)

[6] Martin D. J., Reference Model for Distribution Systems, Working Draft ISO TC184/SC4/WG1, Version 1.1, Document N443, Dec.(1989)

[7] Gerardi M. L., Application Reference Model for Ship Structural Systems, Working Draft, ISO TC 184/SC4/WG1, Version 4.0, June(1990)

[8] Nowacki H., Fernandez-Gonzalez F. et al., Experiences in Neutral Product Data Technology in Shipbuilding, MICAD 93, 12th Conf. on CADCAM,Comp. Graph. and Comp. Aided Technologies, Editions Hermès (1993).

[9] Oehlmann R., Koch T., Application Protocol: Ship Structures, Working Draft, ISO TC184/SC4/WG3, Doc. N302, May(1994)

CALS VISION

A shared data environment created by applying the best technologies, processes, and standards for the development, management, exchange and use of business and technical information among governmental and industrial enterprises.

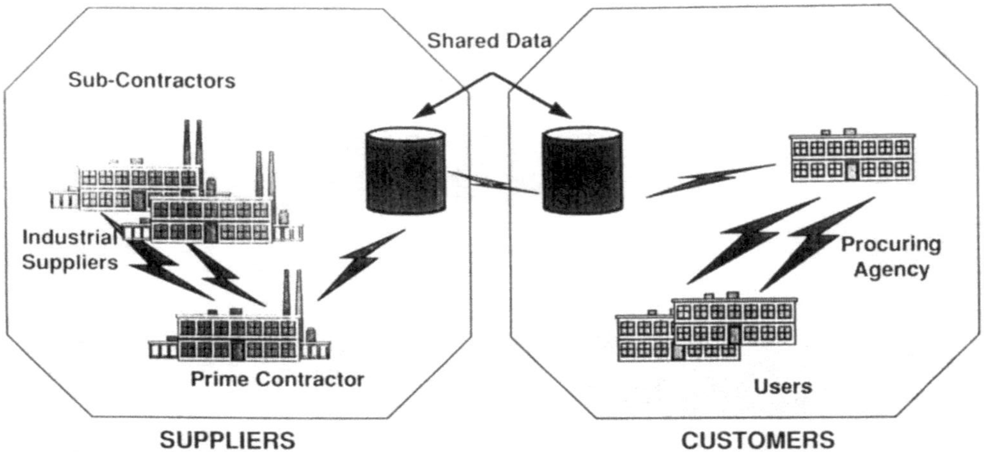

René J. PISTENON

The CALS Initiative: A Global Model for Distributed Data Acquisition

René J. PISTENON

Institute for Research in "Productics" and Logistics, 24 Avenue Prieur de la Côte d' Or, 94114 Arcueil, France

Abstract

CALS, under its new acronym Continuous Acquisition and Life-Cycle Support, is defined as" an Industry and government strategy intended to enable more effective generation exchange, management and use of digital data supporting the life cycle of a product, through the use of international standards, business process change and advanced technology applications". This paper deals with the topic of Data Exchange needed during all the phases of the Acquisition Process of a complex product such as a system implying numerous participants. The design and construction of a space station, for example, tests the limits of our ability to coordinate the decisions and actions of subcontractors with a bureaucracy.

Resume

CALS, avec son nouvel acronyme "acquisition continue et soutien d'un cycle de vie", est défini comme "une stratégie industrielle et gouvernementale orientée vers l'amélioration d'échanges générateurs de la gestion et l'utilisation des données digitales qui sous-tendent le cycle de vie d'un produit, à l'aide de standards internationaux, le changement des processus propres aux affaires et d'utilisation de techniques avancées". Cet article traite du sujet de l'échange de données nécessaires durant toutes les phases du processus d'acquisition d'un "produit complexe" tel qu'un système impliquant de nombreux participants. La conception et la constitution d'une station spatiale, par exemple, permet de tester les limites de notre aptitude à coordonner les décisions et les actions des sous-contractants à l'aide d'une "bureaucratie".

1 Introduction

CALS, under its new acronym Continuous Acquisition and Life-Cycle Support is defined as "an Industry and government strategy intended to enable more effective generation, exchange, management and use of digital data supporting the life cycle of a product, through the use of international standards, business process change and advanced technology applications". [DOD CALS office, January 13, 1994].This paper deals with the topic of Data Exchange needed during all the phases of the Acquisition Process of a complex Product such as a weapon system, a space system, a nuclear power plant, etc.. The design and construction of a space station, for example, tests the limits of our ability to coordinate the decisions and actions of subcontractors with a bureaucracy.

2 ACQUISITION

2.1 The scope

The latest edition of the glossary published by the DoD Defense System Management College (DSMC, 1991) defines Acquisition as covering :

"The conceptualization, initiation, design, development, test, contracting, production, deployment, and logistic support, modification, and disposal of weapon and other systems, supplies, or services (including construction) to satisfy DoD needs, intended for use in or in support of military missions.

The scope of Acquisition proposed by the MIL-STD-881 B (March 25, 1993) is more limited, as it states:

Acquisition commences with the conceptual phase and is completed at the end of the production phase. It excludes all operating and support activities.

In short, the term Acquisition, according to these definitions, covers all the phases of the life cycle of a Product (or System), excluding operational and support activities."

The term Acquisition Program describes a directed, funded effort, dealing with Research, Development, Testing and Evaluation and/or procurement of equipment, with the imperative objective of providing a new capability responding to the needs of a defined mission [in the sense of DoD or any other MoD, the capability will be a similarity capability].

2.2 The needs

The control of the triad --cost, quality, delivery-- extended to the life cycle of a system implies, in the framework of an Acquisition Program:

- A disciplined and rigorous management;
- The justification of all procedures in use;
- The mastery of all information created, exchanged, and used by the different cooperating Government and Industrial partners.

Note 1: The first two issues of rigorous management and procedure justification imply or require that the last one (mastery of all information) be achieved.

Note 2: Each of these requirements must be applied to all phases of the system life cycle.

Note 3: We do not address in this document the organizational impact of the first two issues, "management" and "procedures"

2.3 Fundamental Importance of information

Many government and industrial partners are affected by the Acquisition of complex Systems.

The increasing complexity of products and technologies involved imply, for the different partners, a close collaboration throughout the whole system life cycle: definition of technical needs, design, manufacturing, support of operational readiness conditions, and finally disposal. During all those phases, they will have to exchange a vast amount of information.

2.4 Final objective for the mastery of data

The final objective of the CALS initiative is the incremental integration of data into a single database. This integration will occur throughout the life cycle of a given system. In summary:

To allow harmonization of data exchange during the acquisition phase, the Acquisition Model will need to identify:

- The life-cycle phases concerned by that acquisition. These will include technical activities such as definition studies, manufacturing and logistics studies,
- The major data categories that describe the Product, delivered by CAx software. These categories include contractual, definition, manufacturing, and logistics data.

3 GLOBAL ARCHITECTURE OF AN ACQUISITION SYSTEM DATA MODEL

The aim of this section is to introduce a first level of structuration of the global Acquisition Model and its associated Acquisition Product Database (IAPDB) that we can name System Data Model.

This first level of partitioning is conditioned by :

a. the hierarchical system engineering efforts (System or Product Breakdown Concept)

b. the successive states of a product and associated processes and documents

These two basic partitioning mechanisms are summarized below in paragraphs 1 and 2.

3.1 The System or product breakdown concept

3.1.1. Generalities

The answer to the functional requirements of any System during the early phases of a program consists in proposing subsystem concepts capable of effecting the required functions.

In particular, the System concept defines a system architecture, with its related subsystem breakdown. Subsystem selection allows the development of a precise requirement statement of functional System specifications.

The same approach can be applied at the subsystem level, with a "Subsystem" requirement, and proposed sub-subsystems, capable of achieving each of these subsystem assigned functions.

Each level's requirements allows the partitioning of the system into subsystems and sub-subsystems, down to basic components, yielding what we can call a "System Breakdown Structure" or "Product Breakdown Structure": a product of level n is either a system, sub-system or component.

Note 1: The concepts of systems, subsystems and components are interchangeable. They allow their users, (purchasers, designers and/or manufacturers, constructors or integrators), to consider them as products.

Note 2. The concept of Product, as defined in the RGAero 00040, can be either the grouping of a main system and its support system that constitute the entire deliverable program, or a specific combination of system components, which can be hardware, software, or a combination of both, or the delivery of services connected with the program,

Note 3. The definition given above shows the conceptual similarity between "Product Breakdown Structure" and "Work Breakdown Structure" as it is defined in the following US DoD Instruction 5000.2:

"The WBS shall:

(2) Display [the System] as a product oriented family tree composed of hardware, software, services, data {and facilities (MIL-STD-881)} which results from system engineering efforts during the acquisition of a defense material item."

The only difference between the two is that the RGAero 000 40 definition does not introduce the concept of "data" .

Figure 1, taken from MIL-STD-881, provides an example of the first three breakdown levels for a "Missile" product and can be applied to a Space Program. The complete breakdown structure is the combination of the initially approved breakdown structure and the approved product breakdowns developed by each contractor responsible for the individual product.

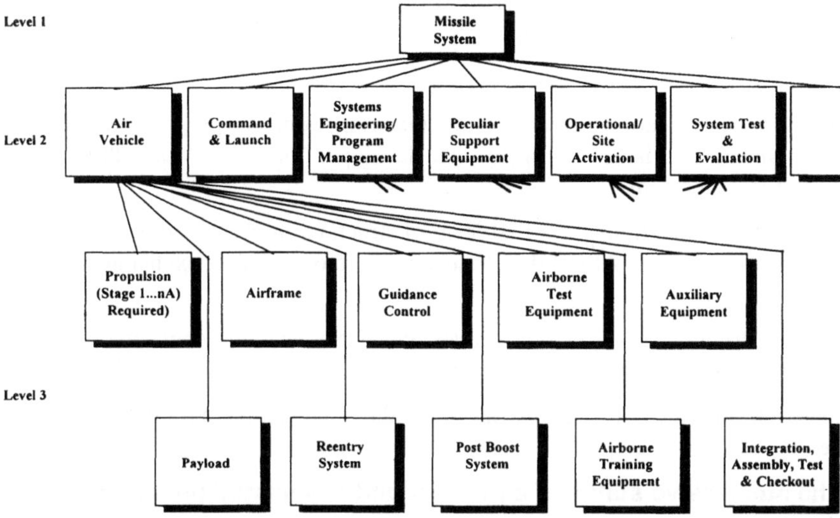

Figure 1. Example of a Product Breakdown : the three first levels of a missile system.

3.1.2. "Tagged" product breakdown structure

The concept of product breakdown structure follows the flow of a progressive product development. It also corresponds, most often implicitly, to the different points of view of the various program partners.

This is illustrated in the following example, which concerns requirements that an nth level product configuration A be composed of B, C, and D.

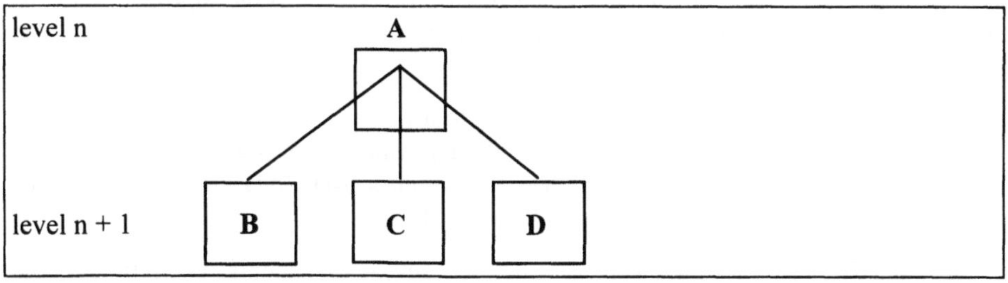

Figure 2. This Breakdown can have:

a functional aspect, from the point of view of the designer : for instance, to ensure the performance of function (A), "Fire Control", it is necessary to have functions (B), "Surveillance, identification and tracking", (C), "Launch control and guiding" and (D), "Communication.

3.1.3. Relationship between product breakdown and underlying data

This relationship is described in different documents and recommendations such as RGAero 000 40 and MIL-STD-881B.

For instance, MIL-STD-881B states:

"The family of specifications and drawings resulting from the progressive steps of systems engineering will conform to the developed program work breakdown structure and its extensions".

This provides the foundations for a first level of structure, or organization, of a global Acquisition Model and its IAPDB data base, and therefore of data exchange between partners.

At this stage, the global Acquisition Model only introduces the concept of System (Product) Breakdown and its tagged extensions.

This approach provides a first partition of the Acquisition Product Integrated database.

3.2 The successive states of a product and associated processes and documents

The "government-industry", "client-server" or "customer-supplier" process which allows one to go from a set of initial product requirement specifications to its manufacturing requires a certain number of studies. Their results constitute the "product data", and will be made explicit in TDP (Technical Data Package). It is these TDP that will be exchanged between the various partners cooperating in an Acquisition program, through exchanges between industrial groups, industry and government, or within government agencies.

3.2.1 The main TDP

The words "TDP" and "documents" used below correspond to the digitized versions of conventional drawings and/or instructions used professionally to describe a given product. To simplify:

- A document (digital data) corresponds to a file or a set of files generated by a CAx system. This document can be the equivalent of a 2-D drawing, such as a mechanical drawing or an electrical diagram, a 3-D CAD model, or a file of numerical control data generated by a CAM processor.

- A TDP will be defined as a structured set of documents, even though the various documents may be of different types. For instance, a Definition TDP will contain :

    - drawings, such as detail drawings, overall drawings and schematic diagrams;

    - technical specifications, such as technical notices and required processes;

GLOBAL ARCHITECTURE of the Product Definition TDP (External Structure)

- **Reference concept**

We will say that the DeTDP A references the DeTDP B if the definition of a product defined by a Definition TDP A requires another product with a definition TDP B.

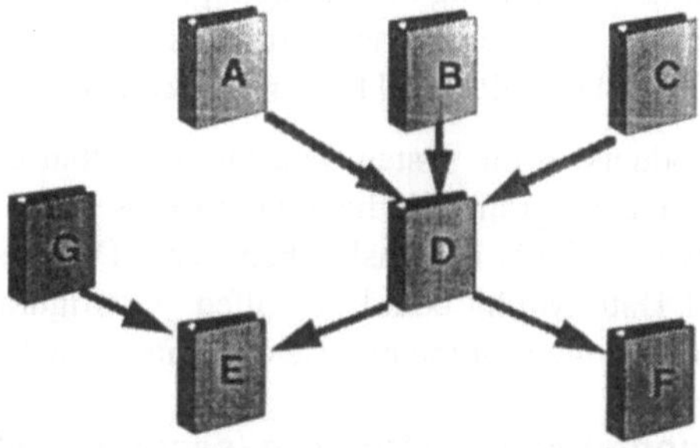

Contents and Structure of a Definition TDP (Internal Structure)

Contents of a hierarchical DeTDP

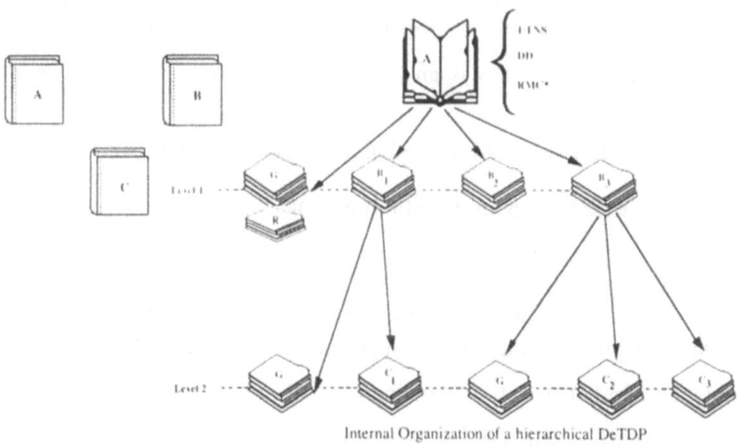

Internal Organization of a hierarchical DeTDP

- lists of components in a nomenclature,
- management data contained in drawing captions.

4 STRUCTURE AND CONTENT OF ACQUISITION TDP

4.1 Introduction: first level of organization for acquisition data

The preceding section has shown that mastery of the product acquisition data implies an initial "system" data organization model. This model relies on the management plan and its base lining, partitioning or organizing these data according to one or more product breakdowns, and for each of these products, according to the major TDP data categories, such as TNS, DeTDP, JoDTDP, and M&CTDP.

For the different products in the system and the TDP that describes these products, one can draw within a diagram all the relationships that ought to be established between TDPs, between TDPs and tasks, between TDPs, users and manufacturers. This global level of Data Model could be called "Configuration Controlled System Data Model". It is independent of the precise contents of each of the TDPs.

4.2 The TDPs: contents and structure (second level of organization)

This second structuring level of the Integrated Product Database, and therefore of the underlying data exchange, is specific to each TDP category, such as TNS and DeTDP.

This is the level where actual technical data conforming to their product data models are exchanged. It will be shown later that these models are more "métier" oriented, i.e. that they are specific to defined technological domains, such as mechanics, hydraulics, electrical, or electronics. This is the case even if a product, as defined by a product breakdown, requires the combination of several métier models.

In general, the reference relation is an approach to managing products that have a partial common definition.

The diagram of the reference relation is a Direct Acyclic Graph (DAG).

4.2.1 Global data attached to each definition TDP

In the Integrated Acquisition Product Database, these data define a Definition TDP and its relation with the other TDPs, such as DeTDP and M&CTDPs, regardless of the contents of this TDP, which can even be empty. To visualize the concept, one could say that these are the data that would appear on the "cover" of the TDP.

Without being comprehensive these data include:

- the TDP identification, such as A and B in the preceding examples. A priori, this identification is not the same as the product identification of the product it describes: while it may be the same, there is no premise that it is and no assumption must be made in this respect. This identification must be unique throughout time,

- The designer responsible for the definition,
- The list of the Reference Manufacturer Codes (RMC) for the product, when manufactured according to the DeTDP,
- The identification of the original Technical Needs Specification (TNS),
- The identification of the TNS that it generates,
- The M&CTDP(s) that it generates,
- etc.

4.2.2 Concept of master and slave documents

In the present state of the art, the exchange data standards explicitly transmit only part of the information necessary for the anticipated future use of these data.

The example of a mechanical part, given in Annex I, shows that the Data Definition Model necessary for its manufacturing, which contains all the data, (dimensions, tolerances, finishing, etc,) and the relationship between these data, such as relationships between spatial geometries, is more an area of research than of "standardization". In this case, it will be necessary to devise several representations for definitions of the same object that conform to several data models, such as 2D drawings and 3D geometry.

5 NECESSARY "CONSISTENCY" OF METIER DATA MODELS OR APPLICATION PROTOCOLS

We have shown that a product DeTDP contains "digital documents" which describe, in particular, the product technical data.

These design data, generated by a CAx software, will be exchanged:

- either between two designers working on the same type of design, i.e. two design engineers provided with different Computer Aided Design (CAD) software having similar function;
- or between designers and manufacturers, for instance between a design engineer and a manufacturing engineer, the former provided with CAD, the latter with CAM software.

For a same technological domain, or "metier", such as mechanics, the data models implicated in the different phases of design, such as preliminary studies, definition studies and manufacturing studies, must be "consistent"; there must be continuity of concepts.

Note: Continuity of abstract concepts exists among the different specialists in the same metier. The Discourse Universe of all mechanical engineers, for instance, is the same. This is in fact exhibited in the industry standards. An example of such standards in mechanics is given by ANSI Y 14.5 for sizing and tolerating mechanical assemblies, where concepts, study methodologies, and (paper-based) representations are consistent.

If one wants to ensure the basic postulate of an IAPDB (Integrated Acquisition Product Database),

"data created once -- data used many times",

the product data models will need to be studied according to technological domain. One will refer more often to product metier-data models. We believe that this is the primary condition to achieve semantically consistent Data Models, and, in the long term, control the information created, exchanged, and used throughout all the phases of the Acquisition Product life cycle.

KEYWORDS

CALS, strategy, exchange, acquisition process, coordination.

Chapter 5

INFORMATION SEEKING STRATEGIES: SELECTIVE ACCESS TO CURRENT INFORMATION ONLINE

INTELLIGENT DOCUMENTATION AND INFORMATION APPRAISAL

Edouard VALENSI

CEDOCAR, Centre de Documentation de l'Armement, 2 bis, rue Lucien Boussotrot, 00460 Paris Armées, France

ABSTRACT

Professional information is based more and more extensively on computer support. One essential objective of documentation is, therefore, to rapidly retrieve data that is "intelligent" or suitable for being quickly integrated into "decision aid support". ST information is only a small fraction of data as compared with industrial property or economic information data. Almost real time "key information" searching capability in a given context is essential. Information must be retrieved and made available, often by tracking through open literature, and using combined search strategies involving a variety of tools and media vectors (text, images, indexes, free references, non-formatted and pre-formatted data).

RESUME

L'information professionnelle tend à passer largement sous support informatisé. Un objectif important de la documentation est donc d'extraire très rapidement des données "intelligentes" ou susceptibles d'être vite intégrées à des "supports d'aide à la décision". Les informations ST constituent une faible part des données par rapport à celles de la propriété industrielle et de l'information économique. La recherche de l'information "clef" est essentielle dans un contexte donné et en temps quasi réel. Elle doit être extraite et rendue disponible, souvent à partir de textes ouverts et par des stratégies de recherche intégrant différents outils physiques et divers porteurs médiatiques d'information (textes, images, indexation, références libres, données sans format ou préformatées).

1 INTRODUCTION

Decision-makers rely more and more on information found in open literature, such as newspapers, professional publications and television clips. The process involves monitoring vast quantities of information and analyzing it to integrate the current state of affairs in the form of updates. There is a constant need for ever more powerful tools to cope with this task.

2 THE EVOLVING DOCUMENTATION SCENE:

Three factors have caused an upheaval in the documentation universe:

- The incredible increase in the power capacities of PCs and microcomputers in general.
- The exponential development of CD-ROM supplies that are revolutionizing the publishing world.
- The development of electronic text files from computer diskettes or from optical recognition processes.

Thus, in our profession, the term "all electronics" is taking on a broader meaning, each year, gradually modifying the scene of documentation. What were formerly only prospective suppositions are now a reality; it is now possible to believe that in ten years' time professional information will be largely memorized on computer or optronic supports and will be retrieved on-line. Our investments and our research programs should be structured with this perspective in mind.

3 PRESS DATABASES: DATA SOURCE IN QUASI REAL TIME

Hand in hand with the broadening of the information market, the number of researchers in science and technology is growing; in parallel, in the field of science the capital of results is becoming richer as studies become fragmented in accordance with the universe, whose diversity we are only beginning to grasp. However, most impressive of all, according to the important information hosts, today scientific and technical information only represents a secondary share of their running expenses. The major part of a solvable market derives from:

- Data bases on industrial property issues
- Data bases on economic information

Beyond that, we must realize that, politically speaking, for an institution like CEDOCAR, all the press data bases in the world are an essential source, making it possible to take instantaneous decisions based on substantially documented files enriched with the most up to date information. World-wide information is not yet available on our desks, but millions of documents can be summoned up in only a few minutes. It is indeed unfortunate that documentary infrastructures are insufficiently developed outside the United States and that Latin, Slavic or Oriental thinking is not sufficiently represented. However, more than half is already available.

Naturally, all is far from perfect; many documents obtained are merely comments with little originality, more or less competent, sometimes partial and might be better replaced by primary information. Reports of studies or original Economy Tables are surely preferable to press communiqués. The difficulty is that, save for speeches by Heads of State, the value of a document is often expressed in inverse proportion to its distribution. You must know how to find the key document among tens of thousands, like the original publication nullifying the patent opposing a firm, or the two in-depth analyses that make it possible to understand the North Korean position in the nuclear controversy. This becomes clear if the information is the key of a contested action, e.g. in law, to be heard, one must produce an original document as proof.

4 FINDING KEY INFORMATION

Whatever the issue, there is no excuse for not seeking out what immediately available information may have been produced by the ingenuity of others, what catalyst will spark your imagination. We are more justified in adopting those first easy or obvious lines of research that led our predecessors to failure. In every area, it is becoming possible to indulge in second degree reasoning. This is true on all levels, up to and including top decision-making levels. It implies, however, that the information be presented in such a way as to be assimilated by the highest authorities.

How do we achieve this? Titles must be grouped together, the assembled documents put in order, the information scanned, then structured, a gradual knowledge of the field acquired, the original documents discovered and a synthesis carried out. In this process, it is important to remember that the solution to a problem only becomes clear to someone prepared to understand or to visualize it. When researching key elements, simple information leads the way for a well-advised informer, and this implies, on the computer scene, software with very specific interrogation formats. A "search target" must be established, even if it is a vague one. This then is our task as informers and these are the actions for which, in the years ahead, we must develop software tools.

5 Multimedia Information and Communication

Over the past few years it has been possible to perfect procedures for arranging documents on the basis of their external characteristics. Documentary software was developed on this foundation, as were the bibliometric analyses indispensable for putting in order hundreds or thousands of references and for bringing forward the broader lines of thought that their authors might have wished to keep hidden. Likewise, we are now confronted by the development of software that will make it possible to handle and analyze from a distance combined CD-ROM information and memorized information in combined form: text, images, characters and texts, indexing and references. We can confidently predict that in five years time, CD-Libraries will have multiplied, causing a complete upheaval in our media library structures and transforming our reading habits.

Conclusion

The ideas presented here are based on the daily experience of a large information centre, the Defense Documentation Center (CEDOCAR), and on its evolution in a rapidly expanding environment, in terms of both methodology and information. In this quickly changing universe, the important question of CEDOCAR's position, as well as that of all information centers, with the appearance of ever more numerous networks and communications, above and beyond interface adaptations, does not imply a basic change in our working methods. The modifications have more to do with easier access from the outside. Our basic objective is still the very broad-based handling of documents to ensure that they yield the significant element or information needed. Salient facts are often far from important, and the evaluation of relevance lies in intelligent awareness to be brought into "man-information" interaction. Properly intelligent tools are then essential for discovering any important information in the shortest and most significant response time. We must remember that the problems encountered in the "enhanced information searches" are usually opposed to those of classical information systems, where the abstracting processes of abstracts or secondary information are always slow. There is systematic opposition between academic needs and intelligent open information requirements. Fast scanning and/or fast browsing in primary and on-line sources are among the biggest challenges for the information systems of the future.

Keywords

Electronic handling, key information, intelligence, strategic data, structuring

TEXTUAL INFORMATION PROCESSING CHALLENGES AND RESEARCH ISSUES

Jean-Marie SÉZÉRAT and Christine COUTRIS

Informatique CDT/RDT, Arcueil, France

ABSTRACT

This document gives an overview of automatic textual information processing, and describes the currently available techniques for processing natural language as an input or as an output for computers. As an output, processing consists of enabling programs to produce relevant documents in good shape. On the other hand, using textual information as an input for programs depends on the system being able automatically to mark and extract relevant information which or buried in the text. In particular, we describe how the structured document technique can help to achieve this double aim, and we present the current lines of research in the field.

RESUME

Ce document fait le point sur les enjeux du traitement automatique de l'information textuelle et présente les réponses techniques actuellement disponibles, tant en sortie qu'en entrée des traitements. En sortie, traiter le langage naturel dans les restitutions consisterait à rendre les programmes capables de produire des documents correctement rédigés et présentés. En entrée des systèmes, l'enjeu est de pouvoir exploiter des documents dans les traitements, ce qui suppose que l'automate puisse repérer et extraire l'information pertinente "enfouie" au sein du texte rédigé. Nous présentons notamment l'apport de la technique des documents structurés pour remplir ce double objectif et évoquons les directions actuelles de la recherche.

1 INTRODUCTION

The amount of information available to us all has risen dramatically in the last few years. There are two technical reasons for this: one is the increasing use of word processors to produce written information, and the other is the expanding telecommunications networks, the spread of the information highways, and the use of communications standards for almost instant communication from one side of the world to the other. Although it is flourishing, this machine-borne information is under-used; recent statistics show that in business, only 7% of information requirements are met by electronic means. The paper habit does not fully explain this under-use of computerized resources. Obviously, the impossibility of talking directly to the machine is one of the obstacles to the more generalized use of the computer in information processing in the broadest sense. This is especially true in the tertiary sector, which is a heavy consumer of textual information: regulations, legislation, texts of agreements, reports, presentations, files, information leaflets, press, etc.

While it is natural for human beings to produce information in natural language and present it in documentary form, it is not the way best suited to machines. The computer can only process information efficiently if it is presented in a very precise and systematic syntactical structure. The two forms, A and B below, provide the same information, but only the second can be directly processed by the machine:

- A: the sentence: "Group X holds 84% of company Y"
- B: the following line, as a table of three columns:

Parent company	*Subsidiary company*	*Share of capital (in %)*
X	*Y*	*84*

Similarly, although they are equivalent in meaning, the computer does not see the two forms C and D below (purely fictitious) as identical:

- C: the sentence: *"If a third child is born, after the 1st January of the current year, the allowance is three times the value set by the finance law, to a limit of 5% of the total declared income"*
- D: the lines of a program (in which the variables have previously been declared):
  IF @NBRINF(n)=3
  *THEN @ALLCE(n)=Min(3*ALL,0.05*@INC(n));*

It may therefore be seen that the information expressed as a written document has to pass through several states for processing by a computer, as follows:

1. Initial written document,
2. Information structured into a machine-readable form,
3. Machine processing,
4. Information from the processing, expressed in coded form,
5. Retrieval edited and presented in a communications medium acceptable to human beings.

The transformations required (encoding between states 1 and 2, formatting between states 4 and 5) are currently carried out mainly by hand. The first level of coding possible for textual information (transformation 1/2) consists in considering it as a character string. In this case, the computer is obviously not able to "understand" the information it is handling, and therefore it is almost impossible for it to process the document using the contents; possible operations include transmission of an identical copy, printing, or some utilities based on character string processing.

Other types of processing need to take account of all or part of the semantics of the document, beyond the simple representational character string. A familiar example would be a business letter. Each country has its own way of presenting these in a standard format. The structure is imposed by the strict conventions relating to this type of letter: the name of the sender is always at the top, the date normally at the top right hand side, etc. Similarly, any written document obeys the conventions of presentation, with a more or less structured format, according to the type of document (chapters have a title which distinguishes them from the rest; paragraphs indicate blocks of text which guide the reader, etc.). Thus the presentation of the documents indicates the internal information structure in the document, a structure linked to a very precise meaning.

One idea underlying the structured document technique consists in exploiting this semantic carrier structure by encoding it. This gives the document a tree structure, whose nodes may be specified, giving a formal structure of the semantics, which is both satisfactory and systematic. This is the technique on which we have focused all our attention. Without it, the text is, for the machine, nothing but a string of disorganized characters, deprived of sense. In the first part, we present current options for retrievals supplied by the processing (stage 4/5) with, in particular, structured documents used to assist the automatic generation of quality documents. The second part covers the possibilities for documents processed as the input to information systems (stage 1/2).

2 IMPROVING RESTORATIONS AFTER PROCESSING

2.1 The challenges

The improvements to the processing output consists in improving the quality of both form and content of the restorations.

Improvement to the quality of the form of the restorations: The processing output of the applications is normally data presented as tables of figures. These results are not immediately usable in this state. They are revised mainly by hand, using currently available automated office tools (spreadsheets, word processing, presentation software) in order to produce an acceptable presentation. This task is costly in terms of the number of operatives it requires and the repeated operations involved. An improvement to this process would be to produce quality documents automatically, which could be used and distributed immediately in their output state.

Improvement to the quality of the content of the restorations: However relevant, a table of figures is "indigestible" material for human beings. When we look at this type of table, we may make simple judgments, such as "this value has increased regularly", "these two values both dropped at the same time last month", etc. This level of reading precedes the in-depth analysis of the variations, and is vital to it.

Moreover, even in the press, much of the material based on numerical information consists simply of a straightforward commentary on tables or graphs. Improving the content of processing restorations would therefore consist in the first place of producing such conventional commentaries automatically, paraphrasing the numerical data in order to guide and assist the reader.

Expected benefits: Improvement in quality of restorations, both as regards form and content, would bring significant benefits, especially for systems intended for decision-makers in business:

- publication of EIS output reports (Executive Information Service),
- improvement of restorations from expert systems or data analyses,
- production of sales literature, summaries or abstracts, for example in order to construct a report or support a decision, etc.

2.2 Available techniques and lines of research

Structuring documents: The technique of structured documents has much to contribute to the production of quality documents. It consists in producing a graphical representation of a logical structure. The generation of a structured document relies on the interpretation of a logical model of the document, in a sense, the grammar of the document. The physical production of the document (printing, for example) then consists in applying a presentation format to the structure, relating to its meaning. This structuring also assists with the physical requirements for restoration: size of page, preferred orientation, line spacing, etc. There are structured document editors available today, as well as a globally approved standard, SGML.

Document handling problems are still to be considered. One of them is the coherence between the presentation format and the document logical structure: the same presentation function should interpret all documents produced from one model

Improvement in structuring of processing: There are techniques available today which allow quality documents to be produced. Automatic generation of documents as the processing output has yet to be resolved; programs should supply the document generators correctly, which assumes the separation of the content generation of the document from the form generator of the document.

Processing must make use of programming paradigms which provide a structured content. One research path consists in exploiting the paradigm of a structured document; the idea being to use the logical model of the document to write the program itself. The structure of the code would then actually be that of the content produced by the processing; this would then provide a structured document, according to a known logical model, which could therefore be used, in particular for quality restorations.

Data processing and presentation of results are presently integrated into the programs. Not only would separating them provide significant improvements in the form of the restorations, it would also facilitate production (encoding) of printing modules, and their maintenance. This area of software expertise is very new. Its development follows that seen already for other areas of knowledge which are now very well-developed, such as coding of human-machine interfaces or mass production systems, for example. Thus, the purposes of the research in this area are to develop a set of compatible and integrated tools for systematic writing of program to separate form from background, and a methodology ensuring that such applications are well-constructed.

Creating comment generators: This framework could be used as the basic software structure for implementing comment generators. Generation of a comment is actually more a problem of presentation than of processing. It consists in describing a systematic "paraphrasing" logic for figures, and therefore assumes a structure supplied by the programs. The greater the detail provided for the semantics of the results, and the description of the operating logic, the more detailed will be the automatically generated comments, giving a progression from one-dimensional comments, to multi-criteria explanations, and even to expert comments.

The main technique used at the moment for text generation itself is that of "text with gaps". Its results are relevant, but the quality of the material produced is limited, since any variation in the text formulation is impossible. One should to build real comment generators, especially with the development of text generation grammars.

3 IMPROVING THE MACHINE'S CAPACITY TO DEAL WITH DOCUMENTS AS PROCESSING INPUT

3.1 The challenge

The challenge in processing text as the system input is to enable the computer to accept documents, and to process them; this involves not only handling the document as a whole, but above all processing the content. The challenge is to detect the relevant information buried in the text. Once it has been accurately detected, the information may be extracted and processed, that is, it will be put through a predefined structure according to the process to be implemented. This predefined structure may be a database format, an executable computer language, or a filtering criterion.

3.2 Techniques currently available

So long as the document is considered as a simple character string, the only technique available to discover the information buried in the text is that of processing the character string itself. This is the feature exploited by spell check programs. They can only mark whether the characters between two blank spaces, or punctuation marks, exist in their wordlist or not. They may thus detect possible spelling mistakes, but not an error of agreement, since this would assume a much more intelligent understanding of sentence meaning. Similarly, the technique for detecting information by **keywords** also consists in recognizing a string of pre-defined characters.

Already complicated by the recognition of proper names, or initials, detection of expression rapidly becomes extremely ambiguous. With the keyword techniques, how is a distinction to be made between "*France participera à* la Conférence de..." (will take part in...) and "IBM *France participera au* capital de X pour 32%.."(will take a share in...)? The use of this technique alone generates a great deal of interference in the detection process. This risk of interference is a serious obstacle to the simple search for information within documents, and it completely prevents automatic processing of documents based on this type of detection. The use of **statistical techniques** is also limited : these techniques mark the multiple appearances of words within a given range. They provide classification, and even filtering, of significant amounts of information, but do not allow processing based on the extraction of relevant information from the text.

Thus, labeling the character strings is by no means synonymous with understanding the meaning. Further progress requires the use of textual information processing techniques. The first of these is **structured documents**. The semantic of the document already allows some ambiguities to be resolved. For example, the structure of a business letter clearly shows whether "J.M. Dupont" is the recipient or the signatory of the letter, and if "Pont l'Evêque" occurs in the address, it is certainly the name of the town, rather than the cheese. The more detail given in the description of the structure, the better the automatic utilization of the information contained in the document. There comes a point, at which the content of a block cannot be described in greater detail; this is the limit of use for this automatic marking technique.

Document handling can be taken further still with automatic language processing (ALP). This method exploits the structure inherent in the language: its grammar. A sentence produced in natural language obeys precise and systematic rules, which are powerful guides to the understanding of meaning, especially in a restricted information area. Current ALP developments represent sentences in the form of graphs, schematizing their syntax. The computer is actually able to identify subject group, verb, direct object group, etc. In graph form, the information may be processed in different ways: for example, on the basis of coincidence with typical sentence graphs, filtering or information extraction may be carried out. Using a database of known concepts, this technique allows certain ambiguities to be resolved: initials and proper nouns identified, finding the references for possessive, demonstrative or relative pronouns; resolution of elliptical language. Use of causality graphs could allow deduction or analysis of text, or even processing generation.

3.3 Applications

Filtering: The tertiary sector uses a great deal of information: financial market operators, risk analyzers, underwriters, etc., all of whom need to react as quickly as possible to events in the economic or financial world. These occupations are supplied with information in real time by the press agencies (Reuter, Agence France Presse, etc.) using the information highway. Filtering this flow allows the broadcast information to be targeted to those in the relevant areas of activity. Putting out a signal (audio or visual) on news relating to a predetermined topic would greatly increase the chance of positive subscriber reaction.

Extraction of information: High power extraction tools may only be constructed using syntactic sentence analysis techniques. Systems able to extract relevant information from a text would have many applications:

- supply of documentary databases; such databases are automatically updated and indexed with greater objectivity than with any other technique, whether the topics are entered manually as required, or with full indexing.
- recovery of archives or composition of bibliographies.
- tracking events or detection of a particular scenario.

This application has still to be developed and event tracking is a current line of research. This consists in detecting events whose approach signals meaning. Example: year N, a large French group, X, takes the majority of shares in a small Canadian company whose activity is subordinate to its own. Six months later, the Canadian company absorbs another American company, with a similar line of business. And so on, until two years later, the French company has control of most of the North-American market concerned after a takeover bid. For those involved in the market, there would be great value in being able to detect in advance group X's strategy for entering the US market.

Automatic application generation: Another application which could arise from automatic comprehension of the content of documents produced would be application generation, which consists in entrusting the computer with everything from understanding the texts through to generating the code required for translating the processing. For example: the processing generation application would be able to analyze the tax codes: it would generate the processing required to accept a tax declaration as an input and generate the notification of tax code as an output. Such developments are presently still in their infancy. It is easy to measure the level of natural language understanding which would be needed for their development.

CONCLUSION

Most people therefore continue to receive the results of the processing, and to address the computer in a very crude form, compared to that of natural language, even though much progress has been made since the early computer languages. Processing input and output both see the **structure** of the expression as crucial. The use of the rules of grammar and syntax leave ambiguities which a human reader can discard spontaneously, by consideration of the context, and by deductions or inferences based on a knowledge of the subject, or even, more generally, by all the "common sense" unconsciously available to most people. Therefore, as well as the structure of the expression, there is the issue of a frame of reference of knowledge which is vital to the understanding of the natural language.

This **frame of reference** must include all the terms used in the subject area, like a glossary. In computer terms this normally implies a database. But database options quickly become inadequate for carrying out complex operations on the text: resolving ambiguities of syntax or elliptical language requires deductions, recovery or inferences based on established information. The technical response consists in using knowledge databases: allowing dynamic discovery of new information, by reproducing reasoning mechanisms, from knowledge stored in the database. This technique enhances the possibilities in this the area a great deal. Therefore, through a combination of structured documents, syntactical analysis and knowledge databases much progress may be hoped for in future in processing textual information.

KEYWORDS

Textual processing, Natural language, Restoration, Comment generator, Structured documents, Filtering, Reference frame.

REFERENCES

[1] Bouchet C., Renard F., Génération Automatique de Commentaires - Application à l'Analyse Financière des Collectivités Locales, Document interne INFORMATIQUE CDC (1994)

[2] Danlos L., The Linguistic Basis of Automatic Generation, Cambridge University Press (1986)

[3] Dillet B., Vichot F., Wolinski F., Automatic Processing of Proper Names in Texts, Association for Computational Linguistics, Dublin (1995)

[4] Fuchs C., Danlos L., Lacheret-Dujour A., Luzzati D., Victorri B., Linguistique et Traitements Automatiques des Langues, Hachette (1993)

[5] Haton J.P., Bases de Données et Bases de Connaissances in Base de Données et Données de Base, Masson (1993)

[6] ISO 8879, Text and Office Systems - Standard Generalized Markup Language (SGML) (1986)

[7] Landau M.C., Sillion F., Vichot F., Exosème: A Document Filtering System Based on Conceptual Graphs, INFORMATIQUE CDC, - ICCS '93 (1993)

[8] Pasquier C., Bible: a System for Design and Management of Context-Controlled Documents, INFORMATIQUE CDC - PODP '92, Washington (1992)

[9] Sabah G., L'Intelligence Artificielle et le Langage, Hermès (1990).

[10] Collectif, Structured Documents, Cambridge University Press (1989).

[11] Stanfill C., Statistical Methods, Artificial Intelligence and Information Retrieval, Text-based Intelligent Systems, Lawrence Erlbaum Associates Publishers (1992)

[12] Vichot F., Wolinski F., Exosème: une application de traitement automatique du Langage, Document interne INFORMATIQUE CDC, (1994)

Semantic Analysis of Textual Information in the Aeronautical Field

***Pascal ANDREI*[1], *Christian KRUMEICH*[2], *Marc CAVAZZA*[3], *Clément PAOLI*[4]**

[1] *Centre d'Etudes Scientifiques de Défense (C.E.S.D.) - AEROSPATIALE, France*
[2] *C.E.S.D. - Madicia, France*
[3] *C.E.S.D. - Thomson C.S.F., France*
[4] *C.E.S.D. - CEDOCAR, 00460 Armées, France*

Abstract

In technological monitoring, the primary difficulty was to obtain access to information. For several years now, indeed since international research networks were interconnected and with the advanced technologies, this barrier has disappeared. This in turn gave rise to another problem, that of rational and oriented exploitation of this well-nigh inexhaustible and diversified stock of information. Against this, the AEROSPATIALE's technological monitoring desk has perfected a semantic analyzer which intelligently extracts themes from automatically unloaded corpuses.

A true expert system with principles closely linked to artificial intelligence, this tool underwent various tests, the results of which are highly convincing. This paper reports on one of the most recent of them, now used experimentally in aeronautic fields.

Resume

Dans les processus de veille technologique, les difficultés ont très longtemps résidé dans les moyens d'accès à l'information. Aujourd'hui, l'apparition de nouvelles technologies de communication a rendu accessibles la recherche et l'acquisition d'information. Cependant, l'exploitation rationnelle et orientée de cette quantité d'information inépuisable et diversifiée reste un problème important. Afin de contrer cette problématique, la cellule de veille technologique de l'Aérospatiale a cherché à mettre en place un analyseur sémantique permettant d'extraire "intelligemment" les thématiques de corpus télédéchargés.

Véritable système expert dont les principes sont étroitement liés à l'intelligence artificielle, cet outil a subit diverses expérimentations au travers de corpus représentatifs issus des forums d'Internet,, de dépêches d'agences de presse et de bases de données commerciales. Cet article rend compte des premières analyses, à la suite d'expérimentations, dans le domaine aéronautique.

1 INTRODUCTION

New technologies such as computer science and electronics and new storage and distribution devices have allowed large databases to expand increasingly. These databases handle huge quantities of information related to science, technology and economy.

Nowadays, an accurate information survey does not encounter many problems for it normally leads to a limited number of answers that can be easily processed without a lot of computing. Conversely, a general purpose information survey often leads to extracting a large number of items and implies querying various databases which cannot be achieved without handling the appropriate tools.

Various and miscellaneous existing information sources have drawn the users into adopting a scientific approach for database processing. A wide range of computerized utilities are now available based on statistical methods for evaluating scientific and technical information.

The collection of these methods is referred to as "library analysis package (or library-metry)". These utilities, first designed for database servers, were reused by various teams that have since developed their own library analysis software programs. These programs are designed to allow another type of use for documentary data through the analysis of fields or information areas by cross-scanning items whereas sequential scanning was carried out previously.

A computerized tool set has been devised to allow quick and dynamic processing of a huge collection of data composed of several topic-based documentary corpuses. However, one should not forget that a number of goals have to be reached among which are the following:

- cutting down information running costs
- checking and validating information
- emphasizing unexpected matches
- emphasizing interest related groups
- emphasizing materials that imply an underlying industry
- specifying the background of a given action (competition level, possible prospects, methods used, state of the art and technique)
- specifying information scattering patterns

Recent information compression and graphic data processing techniques allow one to pinpoint vital information and to highlight some networks based on codes, keywords, sources, etc., notwithstanding attached questions that may arise for non-specialists.

These methods tend to change qualitative analysis, often based on hearsay, for quantitative data on the actual state of science.

However, most of these systems are mainly dedicated to analyzing structured data banks such as scientific data banks (for example syndicated scientific articles) or technical data banks (patents).

These methods have proved efficient regarding such data banks, but they can hardly be used for textual data banks or full length documentation.

If we consider automating the workstation for technological and scientific scanning and strategic data collection these tools may then need the contribution of linguistics capacities for processing this kind of information. This study aims at validating the use of a semantic analyzer for using information from various heterogeneous textual sources.

Our contribution experiments the use of a corpus of information items about aeronautics. This experiment uses an automated system for merging multiple-origin information (press-agency lines, Internet, standard data banks) and a semantic analyzer ensuring reliable data extraction producing developed information within a Technological Scanning and Economic Intelligence framework.

2 SOURCE

The corpus sample to be tested was brought together from three types of information sources.

2.1 Press-agency lines (AFP, AP, REUTER, TASS,...)

Such information was gathered from a station devoted to real time and permanent collection of information materials. All the messages acquired over one month amount to some 100 MB. linguistic processing of data then extracted the data relating to aeronautics.

This extraction led to building a subset of five MB. On this obtained corpus, linguistic processing will be carried out to allow indexing and querying capacities.

2.2 Standard databanks

The data banks chosen for experimental purpose are:

- Jane's Defense Weekly (in English)
- AERO database of the CEDOCAR (in French).

Extraction of these materials was carried out by standard querying of the business information server over a six month period. The 500 items stored are either bibliographies or factual reports. The corpus set up from querying Jane's is made up of open press articles written by journalists specializing in military affairs. The corpus set up from querying the CEDOCAR database is made up of bibliographical reviews.

2.3 Resources from the Internet

To complete our sample corpus we used the news-groups available on the Internet for a number of aeronautics and army related fields, which was very interesting from a methodological point of view regarding the diversity of the available information. We collected about 200 MB of data through the use of five news-groups over a period of one month, from which we stored 10% (+/- 20 MB) based on the extent to which it matched or was relevant to our purposes.

2.4 Merging sources

The automatic merging of the three corpuses thus built finally resulted in a 6000 piece of information textual corpus to work on, i.e., .some 40 MB. The software program that we designed for merging sources is based on the automatic recognition of the different source formats thus allowing a homogeneous layout with which the semantic analyzer can work.

3 SEMANTIC ANALYSIS

3.1 Linguistics approach - Towards a technological scanning oriented tool

Overview of semantic analysis

Storing the main components of the large flows of existing information in order to analyze them has not yet been solved by available systems.

Unlike lexical techniques currently available, the solution we consider assumes that relevant access to stored data is to be carried out only through understanding based indexing.

3.2 The specific linguistics approach

Our approach assumes the following basic principles:

- Memory in the cognitive sense comes before intelligence
- The mental "sayable" implements itself in the Piagetian sense based on acquirements' representations independent of natural languages (Naomics according to B. Pottier)
- The language action implies a dynamic process of comprehension based on acquirements' representations, not on the morphological syntax of speech (final linearization, particular grammar style chosen).
- The intelligence needed for understanding is a symbolic combination based on the acquirements' representations reckoned in speech and the acquirements' representations implied by the wording (the unsaid).

Figure 1: Language action

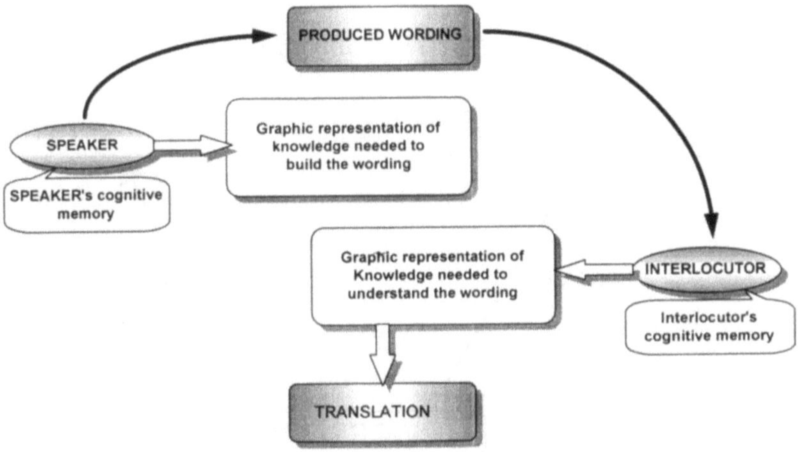

Assuming that both speaker and interlocutor (Fig. 1) are using different languages, how can they understand each other ?

3.2.a Multilingual aspect

Some international bodies solve this problem through the use of an artificial language (for example OTAN language) endorsed by expert committees who state how to word a given situation as regards morphological syntax. These endeavors are not difficult for making choices but are so when final agreement is at stake on how the more or less complex given situation is to be understood.

After a number of artificial language oriented works, we reused the operating outcome obtained when tackling natural language comprehension. As a paradox, a linguist (or psycho-linguist) contributed vital linguistic information from the beginning of the century. Gustave GUILLAUME, who died in 1960 and whose documentary stock is in Canada, used to say repeatedly:

<< **Language** is the "represented" (mental)

and **speech** is the "worded" (use) >>

In Figure 2 tension 1 refers to selection of the representations deemed relevant to the purpose of the speech (tension 2) knowing that the final linearization partly depends on stylistic expression.

Gustave Guillaume operates on the basis of the "universal language" theory (time, space, individuals, designation, etc.) and applies the latter when analyzing specific selections made in Indo-European and ideogram based languages.

Figure 2: Gustave Guillaume Model

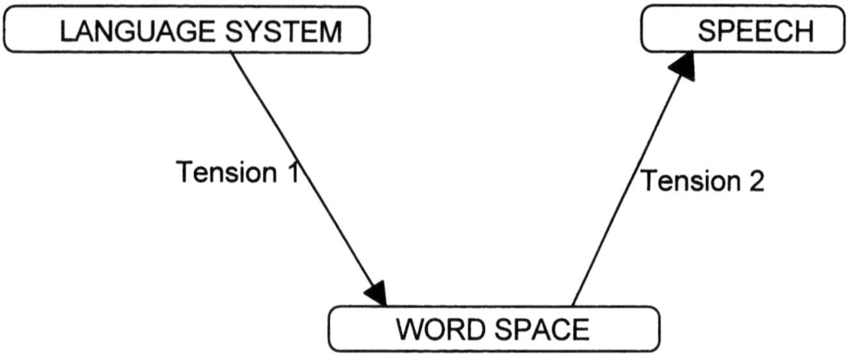

3.2.b The targets

The analyst's goal is to find out which information (language actions) is relevant and may contribute by increasing his knowledge (acquisition of knowledge), this information being presented in any language. It is up to the psycho-linguistic workshop to answer this question as exhaustively as possible.

If we refer to the language action diagram, we can see the following:

Interlocutor	=	Analyst
Cognitive memory	=	Interlocutor's memory of representations and acquirements relevant to the environment studied **and** the acquisition channels ("Reminding" according to R.C. Schank)

3.3 Building an artificial memory

Memory comes before intelligence.

This artificial memory is mainly dynamic and has nothing to do with descriptive data found in a computerized database. It is a dynamic semantic memory and the analyst must be able to interpret, modify and combine any possible representation of this knowledge.

The psycho-linguistic workshop is mainly based on prior encoding of acquirement representations translating the major principles of the analyst's tracking methodology. Then the synonym relationship will complete each major conceptual class for the aeronautics field.

A number of specialized conceptual dictionaries are being set up to carry out a series of capture tests on a large corpus already mentioned. These specialized conceptual dictionaries add to existing linguistic conceptual dictionaries.

3.3.a A conceptual substitution principle.

The conceptual substitution principle allows for any recognized expression (word, idiom, proper noun, straight or customized expression) to be translated into one or more acquirement representation as required in case of uncertainty.

The different filters (acquirement representation combinations) matching the generic queries and event queries added by the analyst are then run according to the resulting translation.

Any text or paragraph matching one or more filter is captured in the acquirements' base. As a result, the analyst processes will lead to modifying some acquirements' representations and upgrading some dictionaries.

3.4 Semantics approach:

The different steps detailed above enable one to build a project for setting up cooperation between both types of techniques so as to offer a tool and a set of methods that facilitate the analysts' work (end users). The linguistic approach we recommend is to be regarded as a genuine innovation. Its main characteristics are the following:

- no morphological syntax analysis
- words are vehicles but have no inner sense
- transmitted information is not necessarily situated in the actual wording
- the meaning of any wording resides in the acquirements' network that the wording implicitly contains
- content analysis can be carried out based only on recognized acquirements' networks
- a knowledge network comprises both the hub language and the semantic networks

Any knowledge representation originates in a set of lexical expressions that can be found in all common languages (French, English, German, Spanish, etc.).

3.5 Methodological principle

The semantic analyzer works by the automatic indexing of specified area materials. A semantic lexicon or conceptual thesaurus has to be worked out for each specific area. Semantic lexicons use a hub language unrelated to national languages. This hub language will be used for indexing bases through processing based on it.

Access to the semantic contents is made through the use of the semantic lexicons that allow linking the words of the text to the matching conceptual meanings. These contents are then used for indexing. The semantic contents can be processed in a number of ways such as bringing them together by specified patterns (description attempts, etc..) or enabling related concepts.

The main benefits from this indexing method are the following:

- knowledge levels classified according to the field
- indexing of implied or explicit research context
- no relationship with natural languages
- indexing speed and real time
- could do automatic summaries
- storage done based on thematic channels

Semantic analysis of the text allows indexing based on ideas, contexts or implied contents.

3.6 The analysis system

The semantic analyzer comprises a semantic indexing engine and a query interpreter both depending on semantic lexicons. It can be described by the three following aspects:

- **Data management**

Indexing, loading and unloading databases, ...

- **Queries or interrogating**

Querying in natural language, selection of a concept from the thesaurus, compounded questioning by Boolean variables, selection of themes, ...

- **Material handling**

Post-semantic-querying documentation building, output, storage, re-indexing documents for subsequent use ...

The man-machine interface uses Windows 3.1 graphical environment. The indexing engine does not depend on the materials language nor on the querying language.

Translation samples

Explosion = Agir-technique-avec_violence, soit ag_tech_vlc
Moyen de locomotion = Etre-technique-se_déplaçant, soit be_tech_dpl
Avion = Etre-technique-se_déplaçant_dans_l'air be_tech_dpl/air

The outcome: The processing aimed at validating the use of a linguistic analysis tool, originally designed to work on geopolitical information materials, applied on more technical areas by creating related semantic conceptual thesauruses. The outcome of the experiment outlines the process possibilities allowed by the use of corpuses both literary (geopolitical) and technological. This validation process permits the consideration of the implementation of full-size systems managing all sides of a strategic information file such as technologies and competition watching, geopolitical background, etc.

Outcome typology: Whatever the source, the process results in obtaining the paragraphs (or notable textual entities as needed) that matched the query variables. These paragraphs are highlighted in the original document, the latter being displayed in its original format.

CONCLUSION

The use of electronic information sources that can be accessed through the various communications networks, whatever their origin and initial display, has been made possible by multiple-domains and multiple-sources "intelligent" automatic analysis systems.

The development of linguistic analysis methods and the integration of the latter into library evaluation tools has produced a comprehensive scanning device for processing structured data banks (science, technique, technology) and textual data banks. The problems stemming from systems operating on textual data banks are the following:

- information is mixed up with a huge quantity of data
- multiple and heterogeneous sources plus the language problem
- query's background unknown to the user

The approach that we recommend, for textual data as well, is to make use of library evaluation tools (lists, matching terms, graphics, etc..) to build a linguistic environment thesaurus.

-This process permits, among other things, reducing the number of problems encountered in establishing the appropriate terminology for the considered applications.

-The device will be based on setting thematic conceptual dictionaries and the conceptual substitution of the documentary material contents.

The system as a whole is devised to specify the scientific, technical and/or economic background of a given question's topic, then once conceptual substitution has been carried out, to analyze automatically the documentary materials and the textual databases contents.

Integrating the different tools detailed above will allow dynamic management of the decision-support systems as well as the competition and technology scanning systems. However these tools are mainly designed for scanning the relevant information and then changing it into developed (or intelligence) information.

It seems obvious that - given the economic wars fought by the technologically advanced countries - validating linguistic utilities in scientific or technical areas should provide a valuable contribution to the strategic information groups and departments within companies and public enterprises.

KEYWORDS

Semantics, aeronautics, textual, watch system, linguistics, cognitive

REFERENCES

[1] Cavazza M., Analyse sémantique du langage naturel par construction de modèles, Thèse de Doctorat - Univ. Paris VII (1991), Modelling based natural language semantic analysis, PhDThesis - Paris VII Univ (1991)

[2] Krumeich C., Indexation sémantique non-lexicale: Application au domaine géopolitique et à la veille technologique, Conférence Ile-Rousse: Les systèmes d'information élaborés, 9-11 juin (1993), Non-lexical Semantic Indexing for Geopolitics and Technological Scanning, Ile-Rousse conference on Developed information systems, 9/11 June (1993)

[3] Paoli C., Krumeich C., Laville F.,Utilisation de l'analyse sémantique comme prolongement des méthodes bibliométriques, Revue Française de Bibliométrie, Vol. N°11, Use of Semantic

Analysis as an Extension to Library Evaluation Methods, French Journal for Library Evaluation, Vol. N°11

[4] Rastier F., Sémantique et recherches cognitives, Paris, Presses Universitaires de France, (1993), Semantics and Cognitive Research, Paris, Presses Universitaires de France (1993)

[5] Sabah G., L'intelligence artificielle et le langage, Vol. 1 et 2, Paris, Hermès, (1989), Artificial Intelligence and the Language, Vol. 1 & 2, Paris, Hermès (1989)

[6] Paoli C., Laville F., Longevialle C., La station d'analyse bibliométrique, Journées ADEST, Paris, 1er juin (1992), The Library Evaluation Station, ADEST days, 1st June (1992)

[7] Paoli C., Billard P., Blanchet P., Longevialle C., Apport de l'analyse factorielle et de la Classification Ascendante Hiérarchique dans l'analyse des banques de données bibliographiques, Colloque SFBA Les journées sur les systèmes d'information élaborée, Ile-Rousse, Sept. (1987), Contribution of Factorial Analysis & Hierarchical Ascending Classification to Bibliographical Data Banks Analysis, Developed Information System Days SFBA Conference, Ile-Rousse, Sept.(1987)

[8] Dou H., Hassanaly P., Quoniam L., La veille technologique et l'information documentaire, Le Documentaliste, vol. 27, n°3 pp 132-141. (1990), Technological Scanning and Documentary Information, Le Documentaliste, vol. 27, n°3 pp 132-141 (1990)

[9] Dou H., Quoniam L., Hassanaly P.,Utilisation de la synergie des ressources de l'analyse de l'information en ligne et hors temps serveur, Colloque SFBA Les journées sur les systèmes d'information élaborée, Ile-Rousse, Sept. (1987), Use of the Synergy of Resources from On-line and Off-line Information Analysis, Developed Information System Days SFBA Conference, Ile-Rousse, Sept.(1987)

[10] Dousset B., Benjamaa T., Une approche interactive de la manipulation, l'analyse et la classification de données bibliométriques par représentations tridimensionnelles, Colloque SFBA Les journées sur les systèmes d'information élaborée, Ile-Rousse, Sept.(1987), A Tri-dimensional Representation Based Interactive Approach for Handling, Analysing and Classifying Libraries Data, Developed Information System Days SFBA Conference, Ile-Rousse, Sept.(1987)

[11] Paoli C., Dionne M.C., Organisation des données bibliographiques pour l'analyse bibliométrique dans La veille technologique, l'information scientifique, technique et industrielle, sous la Direction de H. Desval et H. Dou, Dunod (Ed.) (1992), Structuring Bibliographic Data for Library Evaluation excerpt from Technological Scanning and Scientific, technical and Industrial Information under supervision by H. Desval & H. Dou, Dunod Pub. (1992)

[12] Turner W.A., Chartron G., Laville F., Michelet M., Packaging information for peer review: new co-word analysis techniques excerpt from Handbook of quantitative studies of Science and Technology, A. Van Raad Pub. , Elsevier, North Holland (1988)

[13] Courtial J.P., Laville F., Sigogneau A., Schweighofer M.G., Turner W.A., L'analyse des mots associés, outil de veille scientifique et de pilotage de la recherche dans La veille technologique, l'information scientifique, technique et industrielle, sous la Direction de H.Desval et H. Dou,Dunod (Ed.) (1992)

[14] Co-word Analysis: A Tool for Scientific Scanning & Research Management excerpt from Technological Scanning and Scientific, technical and Industrial Information under supervision by H. Desval & H. Dou, Dunod Pub. (1992)

[15] Schank R.C., Abelson R., Scripts, Plans, Goals and understanding Lawrence Erlbaum Associates Inc., Hillsdale, New Jersey (1981)

[16] Sowa J.F., Conceptual Structures: Information Processing and Mind in Machine Addison Wesley (1984)

[17] Schank R.C., Dynamic Memory, Cambridge University Press, Cambridge and New York (1982)

[18] Guillaume G., Langage et science du langage, Les Presses de l'Université Laval, Québec et Nizet, Paris (1964), Language & Language Science, Les Presses de l'Université Laval, Québec et Nizet, Paris (1964)

[19] Bouton C.P., La signification, Klincksieck, Paris (1979), The Meaning, Klincksieck, Paris (1979)

[20] Merleau-Ponty M., Phénoménologie de la perception, Gallimard, Paris (1955), Perception Phenomenology, Gallimard, Paris (1955)

[21] Pottier B., Linguistique générale Klincksieck, Paris (1979), Basic Linguistics, Klincksieck, Paris (1979)

HYPERTEXT GENERATION OF INDUSTRIAL SPECIFICATION FOR CONTRACTORS IN THE AUTOMOTIVE INDUSTRY

Clement PAOLI[1] *et Wenceslas PETIT*[2]

1 Université de Marne-La-Vallée
2 Université de Marne-La-Vallée, Interface - 6, rue Alsace Lorraine, 76160 Darnétal, France

ABSTRACT

This article describes the problems of extracting technical specifications from CNOMO standards, for use by the PSA and RENAULT groups' suppliers. We introduce a conceptual model featuring a structured description of indexes, based on the HyTime international standard, developed as part of the « Conventions for the application of HyTime ». In this model, the index is considered as a navigational mechanism belonging to a topic map that is independent of the indexed documents. The topic map is connected to the documents with a web of hyperlinks.

RÉSUMÉ

Cet article présente la problématique liée à l'extraction de spécifications techniques du corpus des normes CNOMO, à destination des fournisseurs des groupes PSA et RENAULT. Nous présentons un modèle hypertexte de description des index, basé sur la norme internationale HyTime et développé dans le cadre des « Conventions pour l'application de HyTime ». Ce modèle envisage l'index comme un dispositif navigationnel appartenant à une carte des sujets indépendante du document. La carte des sujets est reliée aux documents par un réseau d'hyperliens.

1 INTRODUCTION

1.1 The CNOMO

The CNOMO (Comité de Normalisation des Moyens de production) is a body set up jointly by the Peugeot-Citroën SA (PSA) and RENAULT car groups. Its functions are to promote the standardization of production facilities and to create issue standards documents and have them implemented. CNOMO standards documents are drawn up and revised by experts from both groups, in technical committees. The objectives are to optimize the efficiency of production facilities, to improve safety, living and working conditions and to meet requirements at the lowest possible overall cost.

1.2 The corpus of CNOMO standards

CNOMO standards are company standards which are often extracted from and supplementary to the highest level French, European or international standards. They deal with topics such as materials from which machines are made, calculations, planning, production and inspection processes, the instruments and tools required for production and inspection, etc. The few groups of problems just mentioned at random show the varied nature of the units involved. They include purposes, actions, properties, know-how, etc. The classifying characteristic consists of organizational links : these are standards governing machine construction for the PSA and Renault car groups.

1.3 Users

Users of these standards falls into two distinct categories : technicians in the PSA and Renault groups and manufacturers of capital equipment. For the latter, the standard is primarily a specifications reference. For example, in a section on hydraulic recommendations, there may be an instruction such as « the return circuit must include an M16x200 pressure takeoff meeting CNOMO standard E05.55.720.N ».So many different documents are referred to in specifications that this considerably increases the number of documents that manufacturers must go through before starting work. In the documents referred to, they first have to sort the sections on the machine to be manufactured. This increases the cost of the machine. In the field of capital equipment, specifications may be classified by process - lathe, milling machine, planing machine, etc. **The purpose of our study is to extract from CNOMO standards the specifications or recommendations applicable to a given process.**

2 The Proposed Solution

Because of the nature of the documents to be processed - the fact that they are non-linear, and have strong connections with documents from another source, and the large number of tables and diagrams needed to understand the standards - we decided to code the documents in SGML.

2.1 SGML

SGML (Standard Generalized Markup Language) is an international standard (ISO 8879:1986) [1] for marking up the logical structures of documents. Tags are inserted to make the sequential, hierarchical and referential structures of the information explicit. Documents coded in SGML comply with generic markup systems known as

DTD (Document Type Definition). Each DTD specifies the permitted structure of a type of document. With SGML, the logical structure of the information can be represented and exchanged independently of hardware and software. The information can also be processed and reused in various ways, some of them unforeseen when the information was created.

2.2 SGML and standards organizations

We approached national and international organizations so that we could work equally on CNOMO standards and other standards of the highest level (NF, ISO, EN, ect). On the European level, the Eurostand SGML 1155 project, to which AFNOR, BSI and DIN contributed, resulted in the European pre-standard ENV 1285:1959 [2] which defines a DTD based on the international standard ISO/IEC TR 9573-11 [3]. We decided to structure our documents according to the simplified DTD of the European standards DTD so that they could be exchanged with these standards organizations.

2.3 Compiling a corpus of work

The corpus of work coded in the SGML format contains classes E05 (Hydraulics), E06 (Pneumatics) and E09 (Pipes and Tap Fitting) of the CNOMO standards, as well as the French, European and international standards referred to in these three classes. All these standards were marked up in the SGML format, in accordance with the simplified DTD. The documents were marked up using Microsoft's « SGML Author for Word ». This tool, which is used with Word 6, allows a Word document to be converted easily into an SGML instance by associating elements of the Word template with elements of the DTD [4].

2.4 HyTime

HyTime [5] is an international standard (ISO/IEC 10744:1992) to represent the structure of hypertext and hypermedia documents. **It defines a series of semantic extensions to SGML, on which it is based. HyTime is a meta-DTD.** This standard contains « architectural forms » which are used to create DTDs. HyTime is suitable for the structured description of links and can be used to describe navigational mechanisms.

2.5 Index model based on HyTime

The proposed model, known as topic relationships is as follows:
Navigational mechanisms such as indexes, tables of contents, glossaries and bibliographical references are mixed independently of the documents to which they belong, in a topic map. This topic map consists of a web of links. Another web of

links, known as an entry map, connects the topic map to the documents to constitute the entries in indexes, glossaries, etc.

The proposed model is based on the hyperlink concept as defined in the HyTime standard. A hyperlink is a relationship between two information objects with a number of properties stated in the form of attributes such as anchorage rules, link ends or traversal rules. The use of hyperlinks allows great flexibility in describing the topic map. Hyperlinks can be nested, one hyperlink can be the anchorage of another. This can serve, for example, to describe in an index, that one term as a sub-entry of another. Lets us suppose that in the topic map, a hyperlink connects « hydraulic cylinder » to cylinder by a hierarchic type relation. To display « hydraulic cylinder » in the index under the rubric « receiver », just consider the previous hyperlink as the anchorage of a new hyperlink connecting it to « receiver ». This new link will then be used as anchorage at the index input.

It is also possible to put hyperlinks of the topic map together in aggregates. For example, with the help of « part of » relation, we can break up a « hydraulic cylinder » into a « cylinder rod », an « air bleed valve », a « shock absorber ». To make each one of these components refer back to « hydraulic cylinder », just reconstitute it by forming an aggregate of the links which are connected to it by the relation « part of ». The formatter will have to interpret this hyperlink and display « Cylinder rod, See Hydraulic cylinder », « Shock absorber, see Hydraulic cylinder ».

The topic map constitutes a constant reference system independent of any particular document. It involves relationships between topics : to say that a « piston rod » is part of a « hydraulic jack » is a « part-whole » relationship independent of the document in which it appears. The compilation of information constitutes a database defining particular scientific, technical or industrial fields.

The topic relationship model was developed by CApH [6] with no explicit reference to similar methods, starting from different points of view but with the same purpose of organizing access to information. To compensate for this, and to show the value of such a generic model, we list specific models which could meet our initial aim of generating technical specifications from a corpus of standards. As far as possible, we choose models from different fields such as linguistics[7] and functional analysis.

3 DEVELOPMENT ENVIRONMENT

Currently, one of the most advanced HyTime engines is TechnoTeacher's HyMinder, which can be used to :

- parse SGML and HyTime documents,

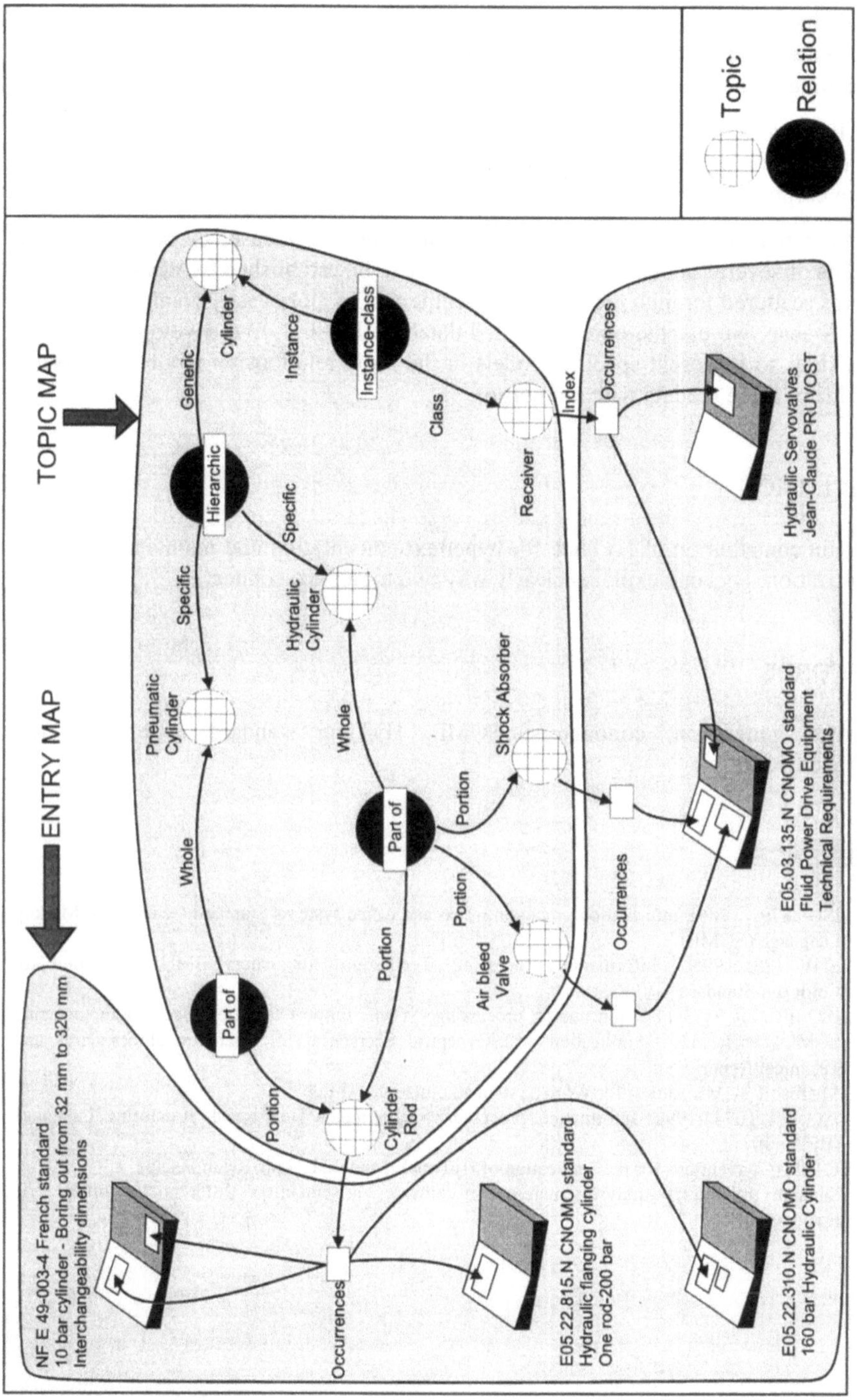

Topic relation map

- create a navigational web of hypermedia objects containing text, images, sound and video,
- navigate in the web and edit, manipulate and validate it,
- generate exchangeable documents in output format according to HyTime.

The HighText's EnLIGHTeN application built around this engine implements the topic navigation model. This tool is used to validate the value of a specific model for a given corpus.

The topic map can either be stored separately, as a database, or be made up of elements of several documents. The topic map can be established by means of simple markups scattered through the different documents. To more easily create and process the topic map, we use the object-oriented database P.O.E.T. We develop tools based on P.O.E.T. to represent specific models in the topic relationship model, and to test the value of these models with our corpus.

CONCLUSION

The main contribution of HyTime for hypertext consultation and maintenance is link semantization. HyTime explains clearly why two topics are connected.

KEYWORDS

Hypertext generation, contractors, SGML, HyTime standard, web, industrial specification.

REFERENCES

[1] ISO 8879 : 1986 : Information processing-Text and office systems-Standard Generalized Markup Language (SGML)

[2] ENV 1285:1995 : Information processing -Techniques for using SGML -Application for European Standard

[3] ISO/IEC TR 9573-11 : Information processing -SGML support facilities- Techniques for using SGML . Partie 11 : Application at ISO central Secretariat for International Standards and Technical Reports.

[4] Microsoft SGML Author for Word System administrator Guide.

[5] ISO/CEI 10744:1992. Information processing -Hypermedia/Time-based Structuring Language (HyTime).

[6] CApH (Conventions for the application of HyTime), Module I,Topic Relationships.

[7] Summary of articles-Analyse sémantique et contexte : la sémantique différentielle unifiée. Eric François.

DATABASE ACCESS : ROLE OF GRAPHICS INTERFACES FOR SELECTING MATERIALS

Jean FAGEBAUME, Jean Pierre CALISTE and Hubert BRABANT

Laboratoire National d'Essais, 1 rue Gaston Boissier, 75015 Paris, France

ABSTRACT

Many information systems make it possible to select materials (metals, plastics, ceramics,...). Such systems are based on a priori knowledge of the subject, and the query language of the database does not help the final user to formulate the "right" question. This paper presents work carried out by MEM and LNE (France) on a new approach combining numerical and graphic information for selecting polymers. Two directions were developed. The first exploits the Minitel terminal combined with a fax output. The second, on PC, is based on a multi-dimensional analysis applied in a space in which materials are represented by points whose coordinates are property dependent. Their distance from an ideal or reference selective point is evaluated.

RESUME

Beaucoup de systèmes d'information permettent la sélection de matériaux (métaux, plastiques, céramiques,...). De tels systèmes sont basés sur la connaissance à priori du sujet, et le langage d'interrogation de la banque de données n'aide pas l'utilisateur final pour formuler la "bonne" question. Ce papier présente le travail fait par MEM et le LNE (France) relatif à une nouvelle approche combinant les informations numériques et graphiques pour la sélection des polymères. La première voie combine les possibilités d'interaction du Minitel et d'un terminal associé de télécopie. La seconde nécessite l'emploi d'un micro ordinateur. Les matériaux sont ici représentes par des points dans un espace géré par leurs propriétés. L'opération de sélection pour une analyse multidimensionnelle est celle de l'évaluation de distances dans cet espace par rapport à un point de référence ou point idéal de sélection.

1 INTRODUCTION

The micro computer and its associated graphic interfaces (Apple, Windows, X Windows,...) have opened up a new dialogue between software and the end user. This increases the possibilities for selecting materials through engineering data.

We shall distinguish between two complementary approaches :

- the development of "user friendly services"
- a really new selection method based on computing resources, data handling and mathematical and graphic representation.

2 Friendly Services

For such new services, only the graphic possibility is used, the structure of the data bank is unchanged. The graphic aspect can be implemented with add on systems similar to fax transmissions.

Figure 1 shows, during a selection of Certified Reference Materials (COMAR-data bank), the histogram of the distribution of materials according to their concentration around an element (or the result of a selection combining several elements together). With this display on the screen, the end user has a good idea of the capability of the data bank to answer his query.

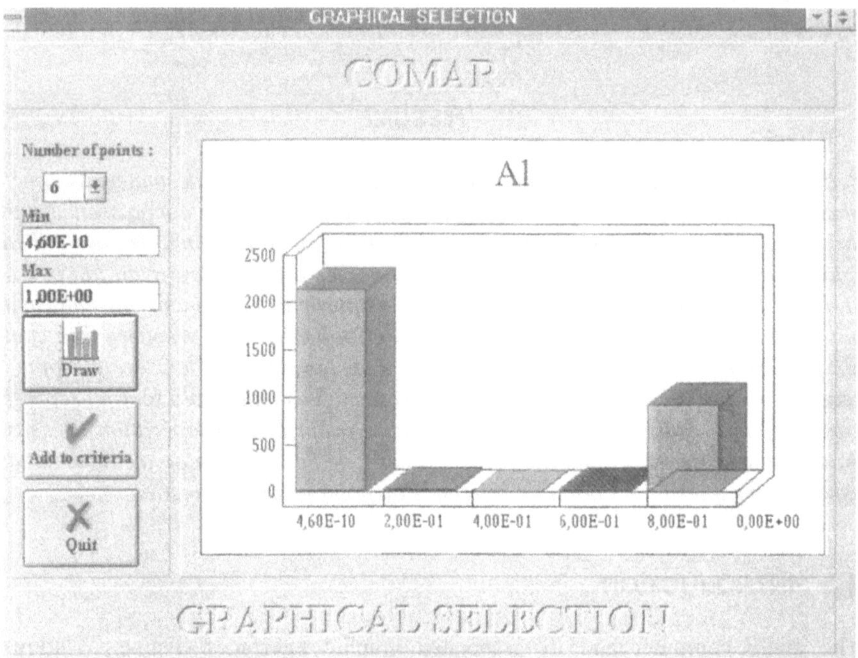

Figure 1. COMAR graphic about the value distribution of A1

Figure 2 shows, for the same data bank (COMAR) the information displayed about one selected Certified Reference Material. The graphic tool allows a more friendly presentation of the data using a "de facto" standardization well known to PC users who require no special training in order to understand and to use it.

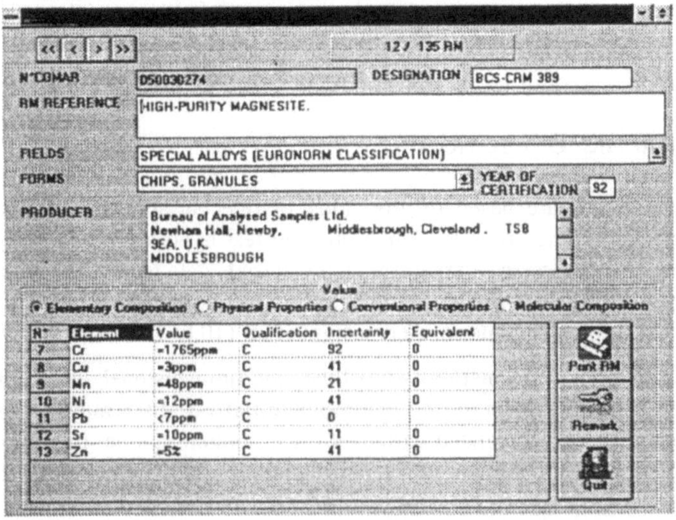

Figure 2. COMAR data information about one CRM selected

Figure 3 is a plane representation of two characteristics (e.g. density and harness), where each point corresponds to a particular polymer (SPAO data bank). By a simple "click" on one point, all factual information can be displayed (identification, producer, values for these characteristics with information concerning test methods, environmental conditions, unit,...).

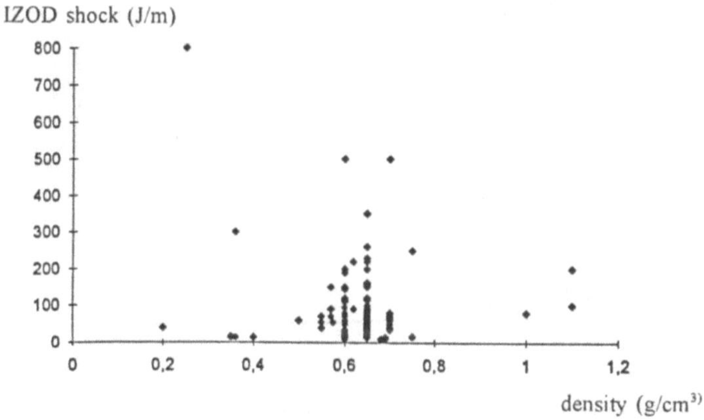

Figure 3. Space distribution of materials according to two properties

This last facility can be considered as a transition between the traditional approach based on menu and a specific interrogation language. The limitation stems from the number of two combinations that can be carried out with all the characteristics included in the data bank.. A multi-dimensional analysis can solve this problem. The second part of this paper describes an original study developed at "Laboratoire National d'Essais" (LNE) for a polymer data bank SPAO.

2 GRAPHIC SELECTION

Classical statistics are based on the study of one variable measured for a limited number of items. They permit estimations and test hypotheses according to a law of probability, e.g. Gaussian. In the case of a materials data bank, the number of items is large (several thousand - 6000 gradations for SPAO), and the number of variables or characteristics for each material can be significant (10 to 100 - 40 for SPAO).

For this kind of information, statistical multidimensional data analyses were developed. They transform the k quantitative initial variables, more or less correlated, to k quantitative non correlated variables. A minimum number of axes (p k) provide sufficient information, allowing for a new and more easily understood representation.

For a first study, we selected only five properties for a limited number of materials (1000). It was possible to distinguish four groups : polyamides (G4), polysulphides (G5), polyesters (G8), copolymers (G14). Figures 3 and 4 show these groups according to axes 1, 2 and 3.

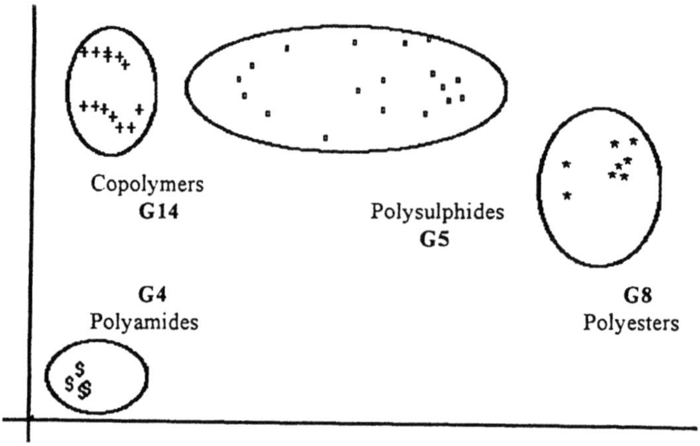

Figure 4. Plane view of the repartition of materials function of five properties (axes 1 and 2 of a multi-component analysis)

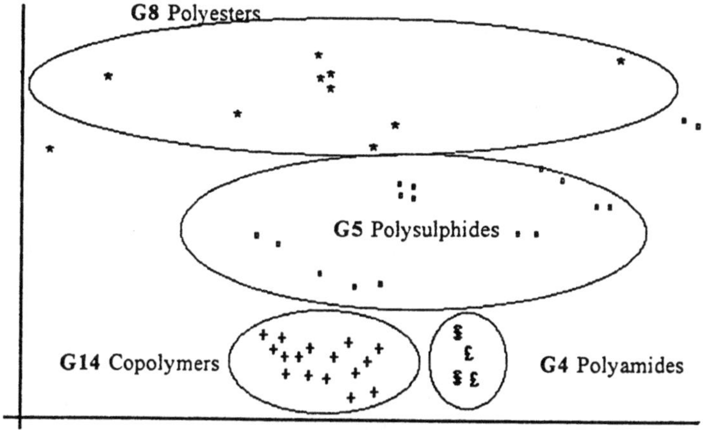

Figure 5. Plane view of the repartition of materials function of five properties (axes 2 and 3 of a multi-component analysis)

These groups were obtained directly from the value property for each material, and with only a two dimensional representation it is possible to have a clear view of the "content" of the data bank. Of course this view can be displayed on the screen of a computer with all the facilities of a graphic interface using "mouse" and a "arrow" in order to select a special point that corresponds to a specific material (see Part 1).

In addition to this representation of all the database material it is possible to display more information in the same representation space such as:

- usage (car industry, packaging,...)
- processing (casting, composite,...)

This information could be optimal and displayed only when requested, and could correspond to different levels of detail (car manufacturing, dash board, hub cap,...) as well as upon user request by an interactive dialogue on the screen.

If the user wants to express a specific need he can define a "typical" material by its properties which correspond to the "best one" for this purpose. The same computing process makes it possible to place an "ideal" point in the cloud of all the points of the material included in the data bank. The point closest to the "ideal point" can be considered as material available for the user needs.

CONCLUSION

The interactive process of selecting material in a factual data bank only begins to use facilities of the graphic interface and the increasing power of microcomputers. The recent development of networks, like INTERNET, opens a numerous opportunities for solving the very difficult problem "how to provide information needed by end users".

KEYWORDS

Information engineering, SGML, hypertext, internet, clustering techniques.

REFERENCES

[1] Lebart L., Morineau A.,Traitement des données statistiques, Dunod-Bordas (1977)
[2] Diday E., Lemaire J., Eléments d'analyse de données , Dunod (1982)
[3] Westbrook J.H., Kaufman J.G., Cverna F., Electronic Access to Factual Materials Information : the state of art, MRS Bulletin, pp 40-48, August(1995)
[4] Rumble J., Computerizing ASTM Tests Methods, ASTM Standardization News, pp 34-37, July(1992)
[5] Newton Crystal N., Computerization of Composites Material Data, ASTM Standardization News, pp 38-41, July(1992)
[6] Marx B., Banques de données Françaises materiaux, Enjeux N° 95, Dec.(1988)

ACQUISITION AND ANALYSIS OF JAPANESE ONLINE INFORMATION

Hélène HAON[1], Jean-Marie DOU[1] and Henri DOU[2]

[1] *Bureau Japon, CEDOCAR, 00460 Armées, France*
[2] *CRRM, Facultés des Sciences de St Jérôme, 13397 Marseille Cedex 20, France*

ABSTRACT

Japan stands for a kind of mystic country where information is hidden and cannot even be seen by non Japanese. This paper tries to clear up the situation of Japanese information for Western users. We have chosen to talk more especially of the on-line Japanese databases. We want to show that Westerners can easily access a large number of Japanese sources of information in English. The point of the discussion will be to quantify what this amount really means compared with the global policy of the Japanese government for databases. Thus we will check what Japanese databases can bring to bibliometric analysis. We present three of the main Japanese databases accessible from Western Europe and the United States.

RESUME

Le Japon passe pour un pays mystique dans lequel l'information est difficilement accessible pour les non japonais. Ce papier essaye d'éclairer la position de l'information japonaise pour les utilisateurs occidentaux. Dans ce but, nous avons choisi de traiter le cas des bases de données japonaises en ligne. Nous voulons montrer que certaines sources d'information japonaises en anglais sont facilement accessible d'Europe. La discussion portera sur la réelle place de cette information dans la politique globale des institutions japonaises sur les bases de données. Nous étudierons ensuite les possibilités d'analyses bibliométriques et statistiques sur les trois principales bases de données japonaises en anglais distribuées en Europe et aux Etats Unis.

1 INTRODUCTION

For Europeans, Japan has always been, and still is, a place at once somehow fascinating and yet impossible to really know. The same applies to Japanese information: numerous stereotypes are in circulation, which prevent a true appreciation of what exists and what is accessible to us. Today, at the dawn of the creation of electronic highways, some still maintain that Japanese information is inaccessible.

In the Japan Bureau of CEDOCAR, it seems to us that it is possible to approach the problem of Japanese information in an entirely different way, particularly when one considers that since 1984 Japan has adopted a policy of distribution of Japanese data banks abroad. This action can be seen as a long-term Japanese plan aimed at occupying a position in the information market on a world scale. Whether it is cars or information, market development policy is often the same.

The purpose of this presentation will therefore be to correct a certain number of false impressions regarding access to Japanese data banks, to show that this access is easier than is generally believed. On the other hand, it is not always as easy as Japanese documentation itself will tell you. We have therefore done a study of the possibilities of access to Japanese data banks on international hosts, and we have then attempted to quantify the real availability of information. Finally, we have carried out a study on the possibilities of bibliometric analyses provided by these data banks, for many still emergent.

2 Databases and Japanese Hosts

In this work, we set out to study Japanese databases produced in English and accessible in the West with standard equipment (micro-computer, modem, communications software). We limited ourselves to information available online (databases on diskettes or CD-ROMs were not studied here despite their importance).

It should be pointed out that when we speak of Japanese databases the source of the information is Japan, and this does not mean data banks dealing with Japan. The particularity of the bases in which we are interested, compared with what we have called "sources of information on Japan", comes from the fact that they consist entirely of documents published by Japanese organizations (publications, companies, universities, scientific associations, etc.). This information is thus primarily intended for the Japanese themselves. The advantage of using this Japanese information, instead of information on Japan, is that we have access to the same raw data as the Japanese.

The information, beyond the criteria of revelation, will therefore only possibly lose its sense in the translation from Japanese to English. One can therefore consider that this information is virtually equivalent to the original information and that it has not undergone any external analysis (American or European) introducing a distortion into the ideas expressed by the authors. This distortion can be considerable for the economic and political press (it is easy to see the differences in points of view between the Japanese, the Americans and the French, for example).

Among all the sources of information on Japan made available to us by the producers and distributors of databases, we have therefore chosen information of Japanese origin, which enables us to work in relation to what the Japanese know and think.

Our research is based on two main works :

- "Databases in Japan 1993", published by the Database Promotion Center (Tokyo),
- "Gale Directory of Online Databases; January 1994", published by KathleenYoung Marcaccio.

These two works do not give the same information on Japanese databases. The varying results of these two works were validated by asking producers and distributors (generally Japanese) for information on their products and their availability in Europe.

Additionally, the source of most of the information from Japan is the press. The market of databases containing press reports is exploding. It is therefore sometimes difficult to monitor the new bases loaded on database hosts and know precisely the list of sources indexed in a database. We have nevertheless drawn up a list of the main Japanese databases in English accessible in the West on international hosts. We will present this list later. However, we would first like to give some details of the development of data banks in Japan and their content, and finally locate the main producers of these data banks.

Certain critical minds maintain that Japanese data banks are not interesting, since they are much smaller than many data bases to be found on the main Western hosts. However, as far as we are concerned this is obviously wrong, because what data banks can give us better information on Japan than Japanese data banks? This criticism probably comes from the fact that Japanese data banks were developed later. We illustrate this comment by Figure 1, which shows, not only the evolution of Japanese databases in Japan, but also the overall evolution of databases accessible in Japan.

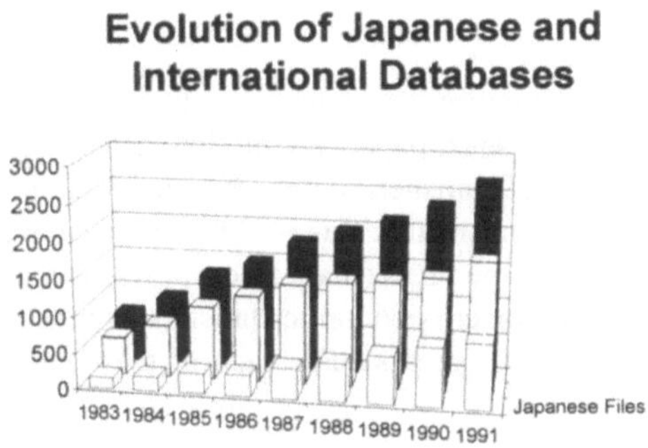

Figure 1. Databases accessible in Japan (Source : Database Promotion Center)

By referring to Japanese literature we have observed that the available information is divided into three major fields:

Press, magazines (35%), Economy, business (48%), Science and technology (20%)

We then observed, and we confirm it later, that the press has a dominant position in Japanese information. To understand this phenomenon, it is sufficient to examine the figures regarding the publication of the principal newspapers in the world. To complete these figures, it should be realized that the specialized press is extremely important in Japan, whether daily (economy, industry, chemistry or electronics) or weekly (cars, aeronautics, electronics, chemistry, etc.).

As regards the producers of Japanese data banks accessible from abroad, we have been able to draw up the following two lists, according to whether the producers were in the public or private sector.

In the public sector, the producing organizations depend on the three main Japanese ministries in which most public research is done in Japan : the Agency for Science and Technology (A.S.T.), the Ministry for Culture and Education (Monbushô) and of course the Ministry for International Trade and Industry (M.I.T.I.).

In fact, NACSIS is hardly accessible from abroad but too important not to be mentioned here. As for producers in the private sector, most notable is the preponderance of press groups, these groups using articles appearing in all of their publications to produce databases, and secondly the emergence of "think tanks"[1].

Until now the publications of this type of organization have often been confidential, but as the number of "think tanks" in Japan has increased very rapidly over the past few years, it is possible that some of them have chosen this new means of distribution to be present on the international information market.

3 Japanese Databases in English Accessible from Abroad

It is not possible to include the entire list of Japanese databases accessible in English with standard equipment in this short paper. People interested may contact us for the details. We arrive at a total of 59 Japanese databases accessible in English in Europe (constant updating of the hosts means that this is a minimum figure). These databases are split into three main sectors: the press (magazines, newspapers, etc.), the economy (stock exchange, economic and macro-economic statistics, company profiles, etc.) and science and technology (research, patents). For each of these categories, we find the following distribution :

Press, magazines: 34, Economy, business: 19, Science & Technology: 6

1 Japanese "think tanks" are organizations which produce studies that usually have a prospective aspect.

In the guide "Databases in Japan 1993" from the Database Promotion Center, we find that 306 Japanese databases are accessible outside Japan. This figure increases every year as shown below in Figure 2.

Date	February 87	July 88	July 89	July 90	July 91	July 92	July 93
Number of databases exported	**28**	**83**	**104**	**155**	**226**	**290**	**306**
Total number of Japanese databases	**425**	**528**	**662**	**808**	**892**	**N.A.**	**N.A.**
% of Japanese databases accessible from abroad	**6.6**	**15.7**	**15.7**	**19.2**	**25.3**	**-**	**-**

After this quantitative view of the available information, we will take a brief qualitative look at these exported databases.

Among the 306 Japanese databases officially available outside Japan, the distribution among the three sectors is as follows :

Press, magazines: 147, Economy, business: 126, Science & Technology: 33

We can therefore compare our results with those of the organization under the control of the M.I.T.I. responsible for promotion of Japanese databases in Japan and abroad (the Database Promotion Center). We have thus observed that among the Japanese databases available in English, the press sector is larger than that of the economy. The economic data on Japan therefore remains largely in Japanese, although accessible in the West.

4 Real Accessibility of Japanese Information

"Databases in Japan 1993" from the D.P.C. gives us figures concerning the databases accessible in Japan: 2,686 databases were distributed in Japan in 1991, of which 1,794 or 66.8% were of foreign origin, and 892, or 33.2%, of Japanese origin. The information available in Japan is therefore overall mainly of foreign origin.

In July 1993 there were also 260 Japanese databases accessible online outside Japan (here we make the distinction between online databases and databases on other media: diskettes, CD-ROMs, etc.). We count only 59 translated into English.

This allows us to calculate that a person who does not know Japanese and is located in Europe can use 59 databases out of 260, or about 23% of the Japanese information distributed online abroad. In fact, this 23% is an average, because there are great differences according to the field. Thus, articles dealing with advanced technology are more often written in English than articles dealing with so-called "traditional" subjects (geology, agro-food, sociology, etc.).

As regards the overall volume of information of Japanese origin distributed online, the percentage is even lower. Use in Europe and in English is reduced to 59 out of 669 bases, or nearly 9%.

For people who speak Japanese outside Japan, the number of bases accessible online is 260 out of a total of 669 Japanese databases, or about 39% of the Japanese information available online.

Summary : **Figure 3.**

Online Japanese Databases		
Total	Distributed abroad	Distributed abroad in English
669	**260**	**59**

Approximation 1991 - 1993

It should be noted that these figures are calculated on the basis that the Europeans mentioned above subscribe to all the available databases. Then the problem arises of the specific needs of each individual for information of Japanese origin depending on their field of activity and the value they give to the information in relation to the cost.

5 Bibliometric Analyses and Japanese Databases

5.1 Bibliometric analyses

Information gathered in electronic form (online databases, CD-ROMs, etc.) can be processed by a variety of data processing resources. Among these various methods, bibliometric analyses allow numerous types of processing of raw information.

These analyses use the structure of bibliographic reference fields from the databases to create statistics of the various elements of the information. Thus the typology of the results will be heavily dependent on the databases used by the data processing software.

CEDOCAR and the SGDN have co-financed a platform, ATLAS (Analysis and Processing of Literature by Automatic and Statistical Methods), for processing information in electronic form. This station integrates numerous information analysis programs from university laboratories (CRRM Marseille, IRIT Toulouse) and also private companies (IBM, Madicia). These programs allow for all possible types of processing: weighting and sorting of documents, statistical analyses, 2- or 3-dimensional representation of data, database creation, re-formatting of files, linguistic analyses, etc.

Our study will mainly use the Dataview program developed at the CRRM of the Science and Technology Faculty of Aix-Marseille III. Using this program we can carry out single-field and multi-field analyses. That is, we can make technical analyses, competitive or otherwise, but also by examining the various bibliographic fields of reference together, obtain technico-economic information, temporal evolution and all sorts of correlation of heterogeneous data.

The results of such analyses are generally represented in the form of 2- or 3-dimensional graphs which visually bring out the important points revealed by bibliometric analyses.

For the most part, the bibliometric analyses were carried out on international databases, INSPEC or WPIL, known worldwide, but the present easy access to information of Japanese origin, and therefore to Japanese databases, opens up new possibilities which we shall attempt to determine here.

To do this, we will examine the two most classic Japanese databases, JICST-E, a scientific and technical database, and JAPIO, an industrial ownership (patents) base, both easily accessible on the STN international host.

5.2 JICST-E

The databases of the JICST (Japan Information Center for Science and Technology) are the classic Japanese scientific databases. All the Japanese publications of these files (originally in Japanese) are translated and combined to form a database in English, JICST-E.

This file is largely inspired, from the structural point of view, by INSPEC (same fields). However, JICST-E has an additional quality: the affiliation of all the authors is present, unlike INSPEC which only retains the address of the first author in its file.

This gives this base a fundamental interest for the study of academic scientific and technical collaborations in Asia. The study of these collaborations is a classic application of bibliometric analyses.

A theme of this study, Japanese/Chinese collaboration since 1993, has been chosen to illustrate the possibilities of analysis on the JICST-E database. We present two typical results of this study: a graph which depicts the main themes through the specific coding of the JICST and a graph of collaboration between Japanese and Chinese researchers.

These two graphs are only a small part of the possible results of such an analysis. In our initial corpus we had over 800 researchers; only 30 are present in this group, which means that the total analysis still brings out numerous disjointed groups.

5.3 JAPIO

The JAPIO database is the English translation of the PATOLIS database produced in Japanese by the "Japan Patent Information Organization". It contains requests for Japanese patents published and not examined: Kokai Tokyo Koho. The patents, covering all possible fields, are indexed in bibliographic form in JAPIO.

The patent database remains very similar to the WPIL (World Patent International Latest) from Derwent. Analyses can thus be carried out in the classic manner. However it has two specific macroscopic classification codes: the JAPIO code and a keyword field. We give two examples of analyses which have generated strategic technico-economic matrices. They reveal the number of patents for each company on a selection of patents relating to heat treatment of steel. Naturally, one can, in other analyses, combine the standard classification codes and specific JAPI codes to allow better segmentation of the accessed data.

CONCLUSION

In conclusion, we hope that this brief presentation will have convinced you of the ease of access to Japanese information, and the possibilities for exploiting this information. We would add two comments to this.

- It must never be forgotten that only 25% of Japanese publications are translated into English. It is therefore extremely beneficial to use the databases.
- As regards exploitation of the results from these data banks with the aid of bibliometry, this is one of our first completed tasks, but we already know that many other types of results will be obtainable shortly.

Note : Contact us by mail or e-mail (jmd@crrm.univ-mrs.fr) for the complete text (with all Figures).

KEYWORDS

Japanese Hosts, JICST, bibliometry, JAPIO.

Chapter 6

COMPLEX DATABASES AND VISUAL ASSISTANCE

A MODEL FOR THE DESCRIPTION OF LARGE DATABASES

Norman SWINDELLS[1] *and Isabel GANCEDO*[2]

Ferroday Limited, 14 Mere Farm Road, Birkenhead L43 9TT UK

ABSTRACT

The problem of describing complex numerical data systems is highlighted. A general model to handle large databases is proposed. The Database, built as a space, is populated with information whose distribution will not be uniform. The space, endowed with more than three dimensions, cannot easily be represented in graphical form but rather as a mathematical model in a metric space with distances attributed to general and specific relations. Hypertext representation using Hypercard is used in the model to simplify examination at different levels, adapted to suit the increasing focus of an inquiry.

[1] Dr. N. Swindells, CEng, FIM, is Managing Director, Ferroday Limited, 14 Mere Farm Road, UK-Birkenhead L43 9TT and a member of the CODATA Task Group for Materials Database Management.

[2] Isabel Gancedo B.Sc is a Consultant to Ferroday Ltd.

RESUME

La description des systèmes de données numériques complexes est traité. Un modèle général pour manipuler de larges bases de données est proposé. La base de données est construite comme un espace de travail, peuplé d'information distribuée de façon non uniforme. Cet espace de plus de trois dimensions est difficile à représenter dans un mode graphique. Il est donc traité comme un modèle mathématique dont l'espace de travail est doté d'une métrique. Ceci permet l'évaluation de distances pour rendre compte de certaines relations. On fait appel à un logiciel Hypertexte, ici Hypercard, pour simplifier les consultations à divers niveaux choisis pour satisfaire le centre croissant de la demande d'un utilisateur.

1 INTRODUCTION

There is currently a great deal of interest in the information industry and its associated market in establishing the quality of information products and services. Quality is defined as satisfying the requirements of the customer at an economical cost which is acceptable to both the supplier and the customer.

The setting of an agreed level for quality implies that there is a description of the product or service on which the customer can base an expectation and that there also exist tests which can measure whether or not this expectation has been reached. An expectation therefore requires an agreed norm against which the product can be evaluated. This norm provides a specification which may be established for a particular situation or may be agreed on a more general basis. The specifications agreed on a general basis are usually called standards when they have the support of some organization with statutory responsibilities. Specifications may in their turn also rely either wholly or in part on relevant standards. A specification for the tests of conformance is also necessary.

In order to establish a reasonable level of expectation and the basis of a quality agreement a method of describing a database is required so that a potential user can understand the scope and purpose of the collection. At the present time there is no agreed method of describing the size and content of databases and various means are used in compilations and directories. The problems with describing bibliographic databases has been discussed by Alkula [1]. The problem is particularly acute in databases providing numerical data for engineering materials. It is not sufficient to list all the materials and all the properties for which provision has been made because data records for materials are invariably incomplete and a simple multiple of the two categories will generate a misleading expectation.

The purpose of this paper is to propose a general method for describing the size and content of databases from which a customer's expectation could be derived and on which a quality agreement could be based. The discussion concentrates on databases of numerical properties for engineering materials but the model should be of more general application.

2 THE DESCRIPTION OF DATABASES

A database should be regarded as a *product* of the processes that created it. There appear to be two main uses for the description of a product:

- To enable a potential customer to determine if the product appears to be relevant to his needs;
- To form the basis of a contract between a supplier and a customer.

The first involves the concepts of expectation by the customer and anticipation of customer needs by the producer or supplier. The second is a compromise between expectation and anticipation so that a working relationship can be established. Alkula only considered the first. This paper is also trying to include the second.

Descriptions are also transformed and adapted for different situations. They may begin as a full specification for the purpose of development. A sub-set of this information could form the basis for sales and marketing literature. There is also a secondary market in providing catalogues and indexes to data systems and all the large on-line hosts, with many different collections of information, provide catalogues of the products they offer for sale. Each of these catalogues and directories impose an internally self-consistent framework to the descriptions of each database to suit their own editorial needs. The transformations into these frameworks of descriptions of data bases designed for different purposes can cause distortions and omissions which can be misleading.

Descriptions of databases have to be sufficient to provide a satisfactory mapping between the expectation of the customer and the delivered form of the product. The most complete description of a database will be the database itself and any other representation will be a model in some form. The critical problem in designing models is to achieve representation without distortion. The development of a satisfactory model requires the use of sets of identifiers which relate to the characteristic features of the original in some way and an associated set of units to establish the magnitudes of any values for the identifiers. The term *identifiers* is used here because descriptors is a term often associated with the labels for the contents of a system. The sets of identifiers and their values will be part of the specification of the system.

2.1 Identifiers of expectation and anticipation

Possible identifiers of expectation from the users side could include:

- relevance (expressed by a mapping relationship);

- importance (expressed by a weighting factor);
- value (the relative importance).

Each of these parameters will be a combination of other identifiers and some desirable features, such as timeliness, may appear in more than one.

Possible identifiers for anticipation provided by the producer could include : purpose, scope, coverage, supporting functions, cost, availability, access, etc. Several User Groups have been active in compiling lists of identifiers for user's expectations and employing these to evaluate databases and database services [2, 3]

The mapping required is between the user's domain of expectation and the producer's domain of anticipation but it can be seen that the identifiers given above relate to different concepts and the expectation of the user has to either form a subset of the anticipation of the producer or the two sets do not overlap at all. Several attempts have been made to develop intelligent retrieval systems to assist user access to complex systems, for example, [4] which imply the existence of a mapping between anticipation and expectation.

2.2 Identifiers of size

Identifiers of size are frequently used in the descriptions of databases. Identifiers which have been used for this purpose include:

- data file size. Typical units: kilobytes, megabytes, gigabytes, number of records, number of items.
- software size (presumably DBMS software). Typical units: kilobytes, megabytes, gigabytes.
- total database size (presumably file size plus software size).
- record size. Typical units: descriptors or data fields per record.
- number of records (total number of records in database).
- completeness (average number of filled fields per record).

Size is clearly thought to be an important component of expectation by the user and of anticipation by the producer since both expect to improve the chances of a satisfactory mapping of requirements if the number of entries is increased. However relevance is a more useful concept for both of them and the identifiers of size given above do not relate to relevance.

What seems to be required is an indication of the number of entities on search and some measure of the extent to which the entity locations are filled - the richness of the collection. The concept of the record, inherited from the early hierarchical DBMS, causes the most problems since it does not apply easily to modern systems such as hypertext and mixed media combinations. As an indicator of expectation it does not do justice to systems which generate further data by calculation. A more general set of identifiers is therefore needed to identify the richness of the contents and to provide an indication of what the customer can expect to find.

2.3 Identifiers of content

Kuhlen and Yetim[5] introduced the concept of an Information Unit (IU) to refer to the different ways of representing basic text constituents, be it as graphs, as tables, as abstracts or as pictures, in a frame-based, automatic hypertext analysis system. An IU is an abstract concept which refers to a semantically coherent piece of text knowledge and which, in combining the different presentation forms, is the main node in a hypertext knowledge base. The IU was defined for this particular case as a 5-tuple:

IU = <ft, ptg, tab, abs, gra>

Where, ft: full text, ptg: partial text graph, tab: tabular presentation of a frame structure, abs: natural language abstract, gra: graphic taken from the full text.

From the relationships built up in their system, Kuhlen and Yetim were also able to determine the relevance of information units with respect to the actual retrieval or dialogue context and the user's intention. This enabled them to define a set of relevant information units (RIU) determined by the relation: share-concept, which exists between the concepts (nodes) of a query and the concepts (nodes) of the pertinent graph. They defined a relevance relation *relevance(Q,G,r)* which determines to what extent the query (Q) overlaps with the semantic representation (G) of an information unit. r represents the measure of the relevance.

3 A GENERAL MODEL

The proposed approach to a general model considers the database as a space which is populated with information distributed throughout the space. This distribution will not be uniform in general and the dimensions of the space will usually be more than three. The space cannot therefore be represented easily in graphical form but must be represented as a mathematical model.

An n-dimensional Euclidean space (R^n) is not a suitable model for a database since such a model implies a space of an infinite number of points and properties of continuity which the information in a database does not possess. The contents of a database can better be seen as isolated points of an n-dimensional lattice: the actually occurring n-tuples of values. It is still possible to define a measure of the database, and its sub-spaces, which will give a size for the existing values only. Also, the positions in the lattice (or combinations of values) which are theoretically possible but do not occur in a specific case, or at a specific stage in the filling of the database, will be disregarded and will not count towards the size of the database.

The mathematical model for such a space is a *metric space*. A metric space consists of a set of points together with a function, called a *metric*, which assigns to each pair of points a none negative real number which is the *distance* between them. The definition of a metric space is arrived at by an abstraction from the properties of the familiar distance in the Euclidean plane or in three-space, and Euclidean spaces are important examples of metric spaces [6].

Formally, a metric space M = {A, d} consists of a non-empty set A together with a map (or function) d: A x A -> **R**. The map is the distance function so that, given x, y in A, the real number d(x,y) will be called the distance between x and y. The elements of A are called the points of the space and d is called a metric on A.

3.1 Measure in one dimension

Let A be the set of values for a given identifier. We can assign an index 0, 1, ...n to each element of A so that $A = \{x_0, x_1, ... x_n\}$.

Note that:

- A is a finite set
- the order in which the indices are assigned to the elements is immaterial. In general, a logical order can be found: an alphabetical list of names, a numerical list of values or a collection of articles, chronologically ordered.

The set A with a function d: A x A -> [0, infinity) and $d(x_n, x_m) = |n - m|$ is a metric space and we can measure the distances between any two elements in the set.

Note :

- d(x, y) = 0 implies x = y M(1)
- d(x, y) = d(y, x) M(2)
- d(x, z) =< d(x, y) + d(y, z) M(3)

The diameter (the biggest distance between any two elements of the set) of a subset B of the metric space A is:

diam B = sup { d(x, y): x, y belong to B}

The function d is the means of measuring the distance and there could be more than one method of measurement, leading to d_i (i=1 to n).

3.2 The general case

Let A', A'', ... A^n be the sets of values of the identifiers defined for a database, and let S be the set of ordered n-tuples of values of the identifiers of the form:

$\mathbf{u} = \{x_i, y_j, ...z_k\}$, where x_i belongs to A', y_j belongs to A", etc.

For **u** and **v** of S define :

$$|\mathbf{v} - \mathbf{u}| = (i_v - i_u)^2 + (j_v - j_u)^2 + ... +(k_v - k_u)^2$$

We define the database space to be the set S with the metric:

$$d(\mathbf{v}, \mathbf{u}) = | \mathbf{v} - \mathbf{u} |$$

The diameter of a subset A of the metric space S is

$$\text{diam } A = \sup \{d(\mathbf{u}, \mathbf{v}): \mathbf{u}, \mathbf{v} \text{ belong to } A\}$$

and the largest diameter can be considered as related to the volume of the space.

This metric also measures the number of entries along the axes of the database. By measuring the diameter of the sub-spaces, or cross-sections, an indication of how the entries relate to each other can be obtained. It is clear that there are also other metrics but the exploration of their properties will be the subject of a further paper.

So in a practical representation of the database, the basis of the model, that is to say the axes of the space, will be identified by the main features of the data system and the volume of the space will be a function of the lengths of these axes given by the number of the characteristic features of that type.

4 THE DESCRIPTION OF A MATERIALS DATA SYSTEM

The information features which could be present in a materials database can be represented by generic identifiers such as:

- names
- number_values
- value_graphs
- value_tables
- text
- images
- parametric_models

To modify the approach of Kuhnen and Yetim it is proposed that these generic data types (classes) should be termed: Information Objects. That is, they are the objects of interest for which the system was created and the features that are processed as a single entity. Each object can have more than one value which will create different instances of the object.

The set of Information Objects, or their instances, represents the basis of the system or the dimension axes of the volume of its information space. The necessary property of the identifiers is that they should also be capable of being represented by a number value to give the size of the space along each dimension.

The whole system is then described by:

- the set of identifiers for the space;
- (AND)
- the set of number values of the identifiers.

The relationships and associations between the identifiers can be represented by sections through the information space. Since each point in the space is either an information object or an instance of an object then each point can have a structure of information associated with it.

5 An Example of a Complex Database

The data system used as an example for the application of this new approach is a hypothetical database for high temperature alloys. The system is assumed to include programs for the statistical evaluation of the data and for the representation of data in graphical forms based on parametric models.

The generic Information Objects providing the first level of specification of such a database would therefore be:

(names, number_values, value_graphs, parametric_models)

The instances of the Information Objects would be, for example :

(alloy_names, test_names, data_category_names, data_field_names, test_results, number_graphs, data_models). These are the identifiers which form the A^n of the metric.

The numerical values of these identifiers could be, for example:

(37, 11, 21, 389, 5820, 4528, 77)

The naming of the objects and their instances would be the responsibility of the database producer but the generic identifiers could be standardized.

6 Visual Presentation of a Description

While the use of metric methods would, in principle, allow for the *calculation* of the characteristics of a database model, it is an advantage to be able to summarize and present the description in a form which is easy to assimilate visually. Two possible presentation methods suggest themselves: graphical plots and hypertext. As an example of a hypertext representation a simple demonstration has been prepared using Hypercard[7].

Figure 1 shows the index of the Hypercard stack which lists the objects included in the specification for this hypothetical database. The hypertext exploration of, say, Data Models, for example, leads first to the index of data models (Fig. 2) and then examination of Creep Models provides the final level of information shown in Figure 3. The benefits of this approach include the capability to nest objects and instances so that a potential user could browse through the model efficiently.

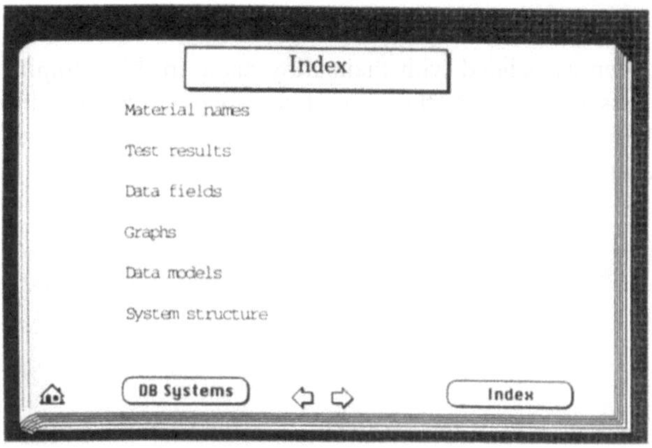

Figure 1.

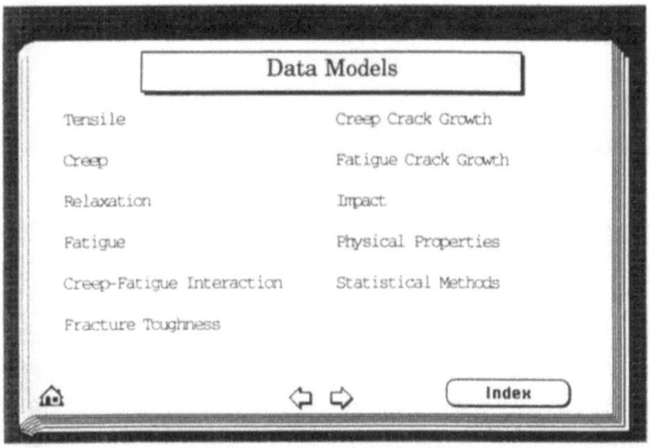

Figure 2.

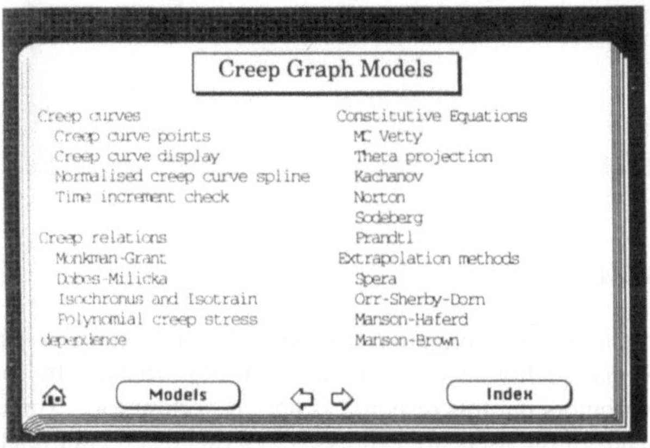

Figure 3.

A cross section through the system is illustrated by Figures 4 and 5. Figure 4 is the end of the link from Test Results in Figure 1.

The results have been associated with their alloy name in this compilation and this was one choice of association, but others are obviously possible. The link from Alloy 800 is shown in Figure 5, which displays the numbers of results from this alloy for the test methods of: tensile, uniaxial creep and creep crack growth.

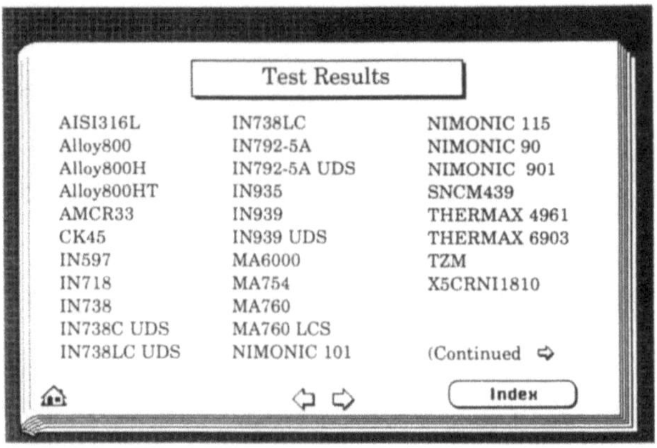

Figure 4.

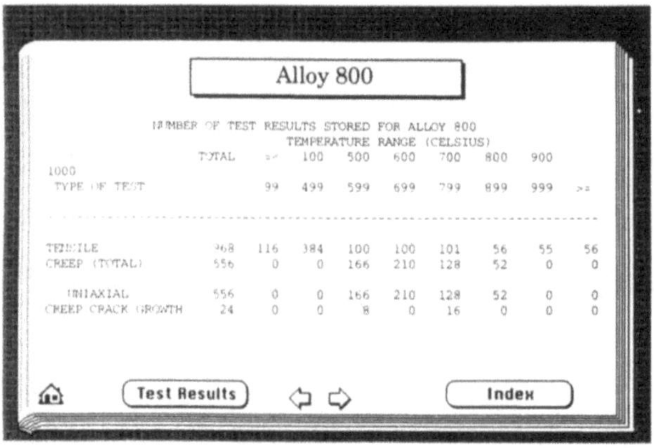

Figure 5.

7 DISCUSSION

The demands of the users for information to support their expectations is being met by the adoption of formal quality management and control procedures by some database producers and distributors but this response is not yet widespread. There is a danger also that the demands of the users are expressed in terms that do not relate directly to the operations of the producers and their distributors. It is clear that there has to be much more use of specifications in order to bridge this gap but equally the specifications have to acknowledge the different domains of interest of the two components of the partnership between customer and supplier.

An understandable and testable description of the database product is central to the requirement for a bridge to be established, and the bridge has to have its foundations in each of the domains.

The model of a complex data base will of necessity itself be complex, and the study of ways to achieve this modeling has only begun. However, there is nothing to be gained from over-simplifying the description if in doing so the richness of the collection is hidden. A model therefore has to be examined at different levels in order to suit the increasing focus of an inquiry. This can be achieved by hypertext methods, using pre-determined links within the model, or by operations which determine sections through the information space at run-time and then display them in some convenient form. This would require the development of an information system for the description, and the potential complexity is beyond the capability of a printed document to display. However there is the need for some aspect of the description to be printable, and the method proposed does have that advantage and provides a starting point for a more detailed investigation.

The method proposed could also form the basis for standardization of the framework for describing databases. The methodology used for such a standard should preferably have a sound intellectual base and be independent of any de facto method.

The approach proposed may not be the only one to use but it does have the virtue of possessing these two desirable attributes. What is clear is that more could be done to provide information about databases in a manner which is more suited to modern systems and by employing methods which use modern modeling concepts and display processes.

CONCLUSION

The problem of describing complex numerical data systems in an effective manner has been highlighted and a possible basis for a more reliable and meaningful description has been proposed. The principle of the method is the use of the mathematical concept of a metric space to model the system. This approach can result in a quantitative description which could be part of a customer agreement based on quality assurance and which could be tested. The procedure could also form the basis for standardization of the method of describing databases which would bring considerable benefits to the information services industry. The method has been demonstrated by using it to describe a hypothetical system for numerical data from engineering alloys tested at elevated temperatures. A more extended investigation using real databases of differing types is obviously required in order to evaluate this method further.

ACKNOWLEDGMENTS

This study originated as a project of the CODATA Task Group for Materials Database Management.

KEYWORDS

Large database, DBMS, identifiers, metric space, distance, visual presentation, materials database model.

REFERENCES

[1] Alkula R., A Framework for Database Descriptions in Proc. 13th Int. Online Information Meeting, Learned Information Ltd, Oxford, pp 77-86 (1989)

[2] Basch R., Measuring the Quality of Data, Database Searcher, 6(8), pp 18-23 (1990)

[3] Juntunen R. et al., Quality Requirements for Databases - Project for Evaluating Finnish Databases, Proc. 15th Online Information Meeting , Learned Information Ltd, Oxford, pp 351-359, (1991).

[4] Reiter M., Improving Online Information Retrieval with an Intelligent Front-End System, Proc. 13th Int. Online Information Meeting, Learned Information, Oxford pp 597 - 604, (1989)

[5] Kuhlen R., Yetim F., Hypertopic - a System for the Automatic Construction of a Hypertext - Base with Intertextual Relations, Proc. 13th Int. Online Information Meeting, Learned Information Ltd, Oxford, pp 257 - 264, (1989).

[6] Sutherland W.A., Introduction to Metrical and Topological Spaces, Oxford University Press (1985)

[7] Hypercard is a trademark of Apple Computer Inc., 20525 Mariani Avenue, Cupertino, CA 95014-6299, U.S.A.

A COOPERATIVE DATABASE SYSTEM BASED ON WORKSPACE HIERARCHY

Didier DONSEZ[1], *Pascal FAUDEMAY*[2] *and Philippe HOMOND*[2]

[1] *Laboratoire HEUDIASYC, Univ. de Compiegne, Centre de Recherche de Progallien, France*
[2] *Laboratoire MASI / UPMC, 4 Place Jussieu, 75252 Paris cedex 05, France*
{faudemay, homond}@masi.ibp.fr

ABSTRACT

Cooperative work enables a user's group to collaborate on the same project. The use of a cooperative database system enables distributed users to cooperate on large data volumes. We modelize the cooperative work by alternative versions which belong to each participant and a merger mechanism which is controlled by a group supervisor. This model is a variant of the distributed architecture model, called the Workspace Model. This model proposes a distributed virtual memory of objects in a hierarchy of servers and clients. We have developed the WEA system, which implements this model.

RESUME

Le travail coopératif permet à un groupe d'utilisateurs de collaborer dans la réalisation d'un projet. L'utilisation d'une base de données coopérative rend possible la coopération d'utilisateurs distants et répartis sur un réseau, sur des volumes de données importants. Nous modélisons ce travail coopératif par des versions alternatives appartenant à chaque utilisateur, et un mécanisme de fusion de versions contrôlé par le superviseur du groupe. Ce modèle est une variante du modèle d'architecture distribuée appelé Modèle des Espaces de Travail, qui permet de gérer une mémoire virtuelle distribuée d'objets dans une hiérarchie de serveurs et de clients. Nous avons développé une implémentation de ce modèle, le système WEA.

1 INTRODUCTION

Cooperative work is an important activity in organizations (companies, laboratories, etc.). This type of activity may be favored by the use of workstation networks. In such a case, group members may communicate between distant workstations, while using computer-based cooperative tools[1,2,3]. Examples of such tools are group editors, group debuggers, etc., e.g. in a group editor several participants update together a same document. Cooperative work may be asynchronous, as when using electronic mail, or synchronous, as for computer-assisted meetings. It may support a small group or a whole organization. Application may be carried out in a single site, with interface replication or in several sites on a common database.

In numerous cases, cooperative work uses a large number of archived data, such as in a CAD or software engineering workshop. Group work must then be supported by the underlying database system [4,5].

In this context, group work may be seen as a specific behavior of embedded and global transactions. It combines a global transaction, called the supervisor, and embedded ones, called the cooperative transactions. More than isolated transactions, we shall consider transaction sequences, called activities. These transactions update database versions. These notions will be explained further.

A cooperative session is then composed of the following steps:

- A user creates a supervisor activity, which defines entry conditions for other participants and decision procedures. Participants join the group.
- A participant of the cooperative group proposes database updates to the supervisor as an alternative version.
- The supervisor informs the other participants of a new proposal. In a first step, this proposal is not known outside the group.
- The supervisor concludes the cooperative work by deciding to set a final version of the database, composed of items from one or several alternative versions. The decision procedure executed by the supervisor is specific to each application and usually implements a consensus within the group.

This model of cooperative work uses known mechanisms from recent database systems. Objects or tuple versions are implemented in several database systems, either relational or object-oriented ones. Embedded transaction mechanisms have also been proposed.

In our approach, cooperative work is supported by a new entity which is both client and server, and which is called the Workspace [6]. This entity supports a new shared distributed architecture for distributed data, which extends the client-server architecture (see also [7]).We first present the Workspace properties which are used for cooperative work. We then describe the main utilized mechanisms. We conclude on the support of cooperative work by the Workspace model and future prospects.

2 THE WORKSPACE MODEL

The Workspace is mainly composed of a database image and of activities. The database image is a set of local objects plus objects included in other, connected workspaces. Activities include service activities which display a view of the database image and application activities, which are composed of transaction sequences. A Workspace can publish its services, it can also subscribe to services from other workspaces. In that case, its database image is the union of local data and of data available through subscribed services. The Workspace may operate in passing mode or in retaining mode. In passing mode, committed updates are immediately copied in the Workspaces where the data come from (the passing Workspace operates as a write-invalidate cache). In the retaining mode, there is a retaining transaction, also called supervisor, which must commit the Workspace updates in order to make them visible to the outside world.

The Workspace implements version mechanisms, which are based on database versions and not on object versions. A database version is derived from a root version, which is itself derived from the initial database state. An object state is the state in the initial database, if it has not been updated since the creation of the new version, and the state in the new version, otherwise. In the case of the components of a composite object, the use of database versions enable knowing the version of components, which would be more difficult with object versions. If a component has been modified since the creation of the new version, its state is that of the new version. Otherwise it is the state of the initial database state or of the root version.

A Workspace can recursively access other Workspaces through a direct acyclic graph of subscriptions. The Workspace hierarchy is well adapted to heterogeneous hardware and networking environments, with various levels of servers. In such environments, various local area networks (LAN) can be connected by one or more wide area networks (WAN). Connection of a LAN to a WAN can be implemented by a connection Workspace, which plays the role of a front-end for this connection. A simple example of a Workspace hierarchy is presented in Figure 1.

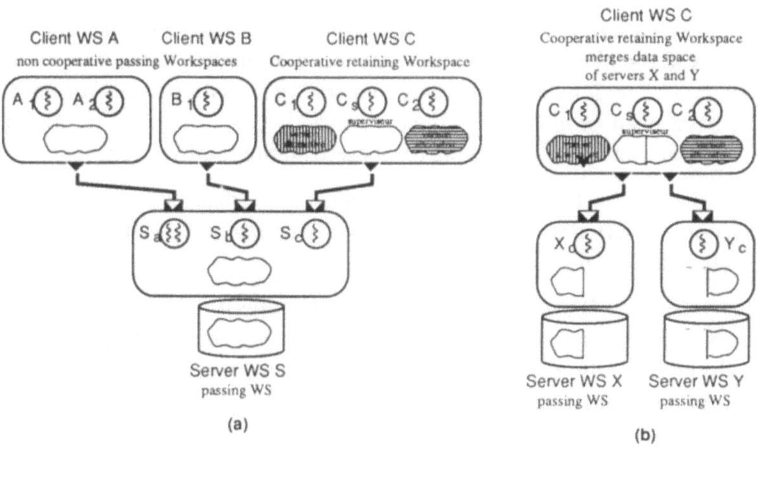

Figure 1. Examples of Workspace connections

Workspace hierarchies are also well adapted to the modeling of cooperative activities in organizations as cooperative subgroups. Each subgroup corresponds to a Workspace. Several subgroups cooperate through a connecting Workspace, which publishes global cooperative services subscribed to by each subgroup. Detailed presentation of cooperating subgroups is outside the scope of this paper.

3 COOPERATION WITHIN THE WORKSPACE

Cooperation mechanisms, which will be described further, do not interpret the semantics of the updates proposed by participants to the cooperative group. Update analysis is left to the designer of the cooperative application who defines the cooperative transactions and the supervisor procedures. Our basic mechanisms define generic operations for generation and update of alternative versions, and communication mechanisms between participants and supervisor.

Our approach to the cooperative work follows a centralized model, in which a participant (or member) of the cooperative group only communicates with the supervisor or by shared access to a version. The cooperative work is modelized by a retaining Workspace, in which each group member is represented by a cooperative activity and the supervisor by a supervisor activity, which is unique in the cooperative group. A shipping service links each cooperative activity with the supervisor activity.

The cooperative work deals with the whole database which is accessed by the Workspace. An update proposal of the cooperative database is presented as an alternative version. A version includes the updates of one or more objects in the database.

The cooperative activity institutes one or more successive cooperative transactions, in order to read the alternative database versions or to update them or propose new ones. The supervisor activity also institutes cooperative activities to read or update versions which have been proposed by cooperative activities. This activity also institutes a "normal" transaction, to update the database with the final version.

Cooperative transactions are different from other transactions (such as application transactions or services), as they never directly update the database. Proposed updates are made in the database version.

In the cooperative Workspace, a version is the property of one of the group members. There is a final version, which is copied back into the database, and which is the supervisor's property. Committed updates of a version are seen by the supervisor and usually by cooperative transactions.

A version is usually updated by the following transactions belonging to the same owner. The version owner and the supervisor may possibly decide that a version is stabilized, in which case it can no longer be updated. The succeeding version of its owner is then derived from this stable version. The root version is considered as stable and visible to all. If a group member wants to keep a version at a given date, he must ask the supervisor for the stabilization of this version.

The cooperative transaction ends either by the validation or by the abortion of the proposed updates. The validation stores the updates in the current alternative version of the group member or creates a new derived version. After the validation of the cooperative version, the corresponding cooperative activity informs the supervisor activity of the proposed modifications. The supervisor checks the modifications and notifies them to the other cooperative activities.

3.1 Supervisor activity

In the cooperative group, the supervisor activity is in charge of :

- the moderation of modified or derived versions and their notification to the participants. The supervisor activity may ask a group member to modify its version before making it visible to the other group members.
- the compilation of the final version according to some decision procedure. In general this procedure includes a notification to other participants of the modifications proposed in a version by each group member, a vote (choice of a note) on each modification, together with vote explanations which are annotations visible by all group members, a decision, which is the choice of a version for each cooperation granule (e.g. by majority vote), and the publication of the result to group members and outside the group.

Notification is sent to the other participants of the name and the author of a modified version (possibly together with modification date), references of all modified objects inside this update (and possibly the localization of the update inside each modified object). The notification precision depends on the grain of the objects which compose the version.

Choice of a cooperation grain is important for the compilation of the final version based on the participants' versions, e.g. if the cooperative work is the writing of a document, the final document may be composed of sections from the different versions, or of paragraphs from the different versions, etc.

The compilation of the final version by the supervisor enables flexibility in the use of the cooperation grain. For example, it is possible to notify a modification from character i to character j in object O, or to insert in the final version a modification characterized by an object name and by a character's range of indices.

Therefore the role of the cooperation grain is to facilitate the phase of negotiation and vote. For instance, if characters i to k of an update belong to a granule, and characters k to j belong to the next granule, there will be two independent votes on the two granules (which may belong to consecutive paragraphs).

3.2 Example

Let us illustrate these mechanisms with the example of Figure 1. The cooperative database is first composed of white objects. The cooperative activity C1 has already proposed an alternative version VA1. In our figure, objects Spades, Clubs and Diamonds represent composite objects which serve as the cooperation granule of the application. A composite object is modified if one of its components is updated.

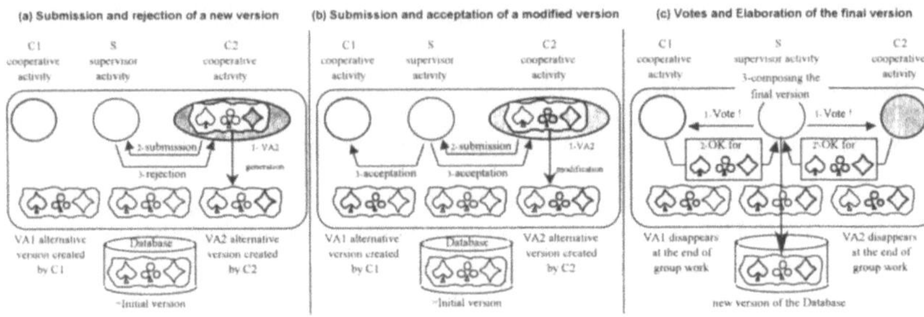

Figure 2. Example of cooperative session

a. Cooperative activity C2 creates a new version called VA2 and submits it to the supervisor activity S. S acts as a moderator and refuses to transmit VA2 to the group.

b. Activity C2 modifies version VA2 (object Clubs is modified), and submits it again to the supervisor activity, which accepts it and transmits VA2 to the group.

c. (1) Activity S starts the consultation of cooperative activities in order to compose the final version.

(2) Cooperative activities answer by voting. (3) Activity S receives the vote results, and decides to compose the final version with updates which have received unanimous votes such as gray Spades object and hatched Diamonds object. The Clubs object is therefore not selected for the final version. The supervisor activity ends the job of groups C1 and C2 by validating this version of the database. Then alternative versions can be eliminated.

Cooperative work and cooperative steps

The cooperative work is a succession of cooperative steps. A cooperative step includes proposals of alternative versions and composition of the final version by the supervisor activity. This final version is then used as the initial version for the next step. The cooperative step can end by the dropping of alternative versions produced during this step. It can also be desirable to keep the previous alternative versions if consensus in the previous step has not been reached on all objects. However, if alternative versions are implemented as differences vs. the current root version, it may be difficult to keep the alternative versions of the previous step.

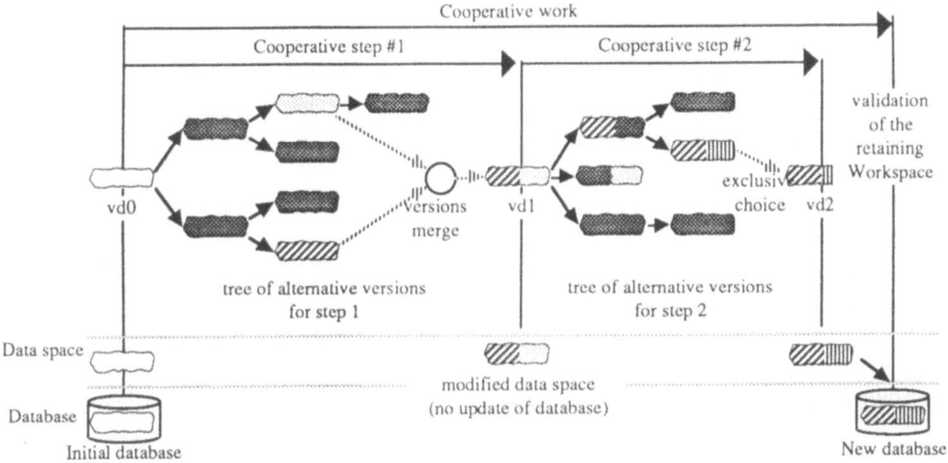

Figure 3. Cooperative steps

Figure 3 illustrates a succession of cooperative steps. The group work begins on an initial database image (in white), which corresponds to version vd0. The end of the first step produces a final version vd1, by merging several versions of vd0. The second step begins with vd1 as an initial version. Alternative versions of vd0 are dropped. A second step ends by the composition of final version vd2 based on a single alternative version. At last, the cooperative work ends by the validation of the retaining Workspace, which validates the database modifications resulting from vd2.

CONCLUSION

The Workspace model displays a natural way of managing cooperative activities which update a persistent distributed database. The model captures the basic activity mechanisms, while leaving the application semantics and the detailed negotiation mechanisms to the cooperative activities and to the supervisor.

This approach appears to be well adapted to synchronous as well as asynchronous operation modes, in various dimension groups. It also keeps the properties which directly result from the model, such as the ability to combine activities from several groups and individual activities (Fig. 1) and the adaptation to many network configurations [Carey 94]. Therefore it is an alternative to approaches based on application sharing. Its properties also include the access to cooperative services by members of distant groups, resistance to failures occurring during a group session, and the ability to structure the group into several sub-group levels. These properties will be the subject of future papers.

We have implemented the Workspace model on a Sparc workstation, under SunOS 4.& 5. This implementation is compatible with cooperative activities. Future tasks of the project include the implementation of libraries for the design of supervisors and cooperative activities, experimentation on real working groups and the evolution towards multimedia cooperative activities.

KEYWORDS

Cooperative Work, Object Oriented Databases, Distributed Systems, Workspace Model, Distributed Virtual Memory, Versions.

REFERENCES

[1] Palmer J.D., Fields N.A., Guest Editors' Introduction: Computer Supported Cooperative Work, IEEE Computer, vol. 27, N° 5, May (1994)

[2] Reddy R., Srinivas K., Jagannathan V., Karinthi R., Guest Editors' Introduction : Computer Support for Concurrent Engineering, Computer, vol. 26, No 1, 12-16, Jan.(1993)

[3] Ellis C.A., Gibbs S.J., Rein G.L., Groupware : Some Issues and Experiences, Communications of the ACM, vol. 34, No 1, 38-58, Jan.(1991)

[4] Kamita T., Ichimura S., Okada K.I., Matsushita Y., A Database Architecture and Version Control for Group Work, Proc. 27th Hawaii Int'l Conf. on System Sciences (HICSS-27), vol. III, Maui, Hawaii, 438-447, Jan.(1994)

[5] Kirsche T., Reinwald B., Wedekind H., Processing Dynamic Interactions in Cooperative Databases, Proc. 27th Hawaii Int'l Conf. on System Sciences (HICSS-27), vol. IV, Maui, Hawaii, 733-742, Jan.(1994)

[6] Donsez D., Faudemay P., Homond P., WEA, A Distributed Object Based on a Workspace Hierarchy, Proc. IFIP WG 10.3 Working Conference on Applications in Parallel and Distributed Computing, IFIP Transactions A-44, Caracas, Venezuela, North-Holland, 247-256, April (1994).

[7] Carey M. et al., Shoring Up Persistent Applications, ACM SIGMOD, Proc. Int'l Conf., Minneapolis, Minnesota, USA, Mai 1994, SIGMOD Record, vol. 23, No 2, 383-394, June (1994)

CASE-BASED REASONING MANAGEMENT OF A STRUCTURAL DATABASE

Jacques-Emile DUBOIS and Belhadri MESSABIH

Institut de Topologie et de Dynamique des Systèmes de l'Université de Paris VII, associé au CNRS, 1, rue Guy de la Brosse, 75005 Paris, France

ABSTRACT

DBMS systems and databases in chemistry, whether for retrieval information or retrieval strategy, are usually limited by their procedural reasoning tools. The limitations of usual systems are often caused by a lack of flexibility due to rigid structuration of the search space, the choice of a given logic and, very often, the fuzzy nature of some of the basic state of space blocks. We are currently investigating the power of analogical search and retrieval tools with an eye to developing a flexible system through the "case history" strategy and the use of "case memory" storing potential similar features chosen by users. The context adaptability parameters, ruling the DBMS reorganization should belong to numerous spaces such as "topology, metrics and physical properties".

RESUME

Les systèmes actuels ne supportent usuellement qu'un seul mode de raisonnement: "le raisonnement procédural". Des progrès importants impliquent différents modes de raisonnement en IA (déductif, inductif, abductif). Cependant, à cause de la prématurité "technique" d'une telle intégration dans un même système, nous proposons une approche basée sur le raisonnement par analogie "le raisonnement à partir de cas" autour d'un système de recherche structurale floue. L'apprentissage à long terme à partir d'une mémoire des sessions d'interrogation permet une adaptation dynamique des métaparamètres stratégiques de recherche et organisationnels de la base de données à un contexte d'utilisation. Dans le prototype développé sous UNIX nous acquerrons une grande flexibilité en combinant les outils souples du flou générique du système DARC à un programme d'apprentissage par induction à partir de "cas similaires comparés à différents niveaux" (topologie, métrique, propriété).

1 INTRODUCTION

The concept of an adaptive and flexible system, capable of internal reorganization according to the problems to be solved, was oriented by our knowledge of an original system of molecular chemistry developed from the DARC system and producing data banks and various expert systems. We note that "structural databases" today constitute the "heart" of chemical information systems. These were developed by successive generations. The first generation was described by fragmented models and handled by low level languages (logical combination of fragments). It was an application of engineering to documentation techniques. Because they divide compounds into fragments to adapt them to informatic handling, these models distort and reduce the semantics of the structural information. The second generation, based on the DARC topological model proposed by J.-E. Dubois had a more intensive scientific development. Chemical structures are here perceived as oriented graphs implemented as tables and handled by graphic language. This generation, with its important scientific activity, was industrialized (EURECAS and CAS). Today, these systems currently include 11 million compounds available on line by substructure. Moreover, certain systems of patent banks also derive from it. These highly performing systems, endowed with generic tools, primarily use procedural reasoning. Most of the derived molecular expert systems are only partially linked to these large databases oriented and optimized for well defined documentary requirements. To go beyond their limitations, it is well known that new systems must dispose of a "multiform reasoning mechanism".

Multiform Reasoning and Data Handling: Like the rich variety of human thought mechanisms, there are several types of reasoning in Artificial Intelligence. The main problem involved in formalizing reasoning is to succeed in conducting a series of operations on symbolic entities while preserving their veracity and the coherence of deduced conclusions. Chemistry is privileged by its symbolism of structures and families. Mathematical logic is also a perfect example wherein formulation is an essential aspect. Roughly, logical reasoning can be divided into three categories: deductive reasoning in which valid conclusions are deduced from a set of premises, inductive reasoning where, on the contrary, premises are generalized and adductive reasoning where plausible reasons are linked to a set of premises often based on limited relations. Many languages and systems are founded on deductive reasoning. This type of logic has serious advantages for building a knowledge based system. The reasoning rules are syntactically and semantically defined; they are moreover simple and easy to use. However, this logic contains intrinsic limitations that prevent it from efficiently dealing with problems of great importance such as analogy, the uncertain, dealing with time, questioning conclusions, etc.

Learning by analogy: Granted the "technical" premature nature of integrating all these modes of reasoning in one system, we propose in our study to integrate reasoning by analogy whose application aspects will be those of a search system of chemical information where analogy searches are carried out on the level of substructures. Our reasoning by analogy makes use of prior cases and can thereby be likened to learning techniques, even if our ambition is for analysis to use and integrate different logics (deduction, induction, abduction).

2 REASONING BY ANALOGY AND INFORMATION SEARCH SYSTEMS

Analogy plays an important part in human daily and professional activities: it intervenes in learning, in decision making, argumentation, explanation, understanding language, etc. More specifically, it is closely related to the processes of memorization and the use of stored information. For this reason, analogy can be a source of inspiration and of carrying out "intelligent" systems for managing large data and knowledge bases. It can be used in two ways: to help a user to find the knowledge corresponding to his need and/or to help improve the quality of the database management system. "Case-based Reasoning", derived from analogy reasoning, is a promising approach for the practical realization of such systems. The idea is to start from previously encountered "experiences and cases" in order to solve new problems. In case-based reasoning, the handling of a new problem is intuitively linked to the learning mechanism, thereby making it possible to assimilate this problem and to work out a learning mechanism for some later use[3]. Figure 1 below illustrates the process of case -based reasoning and learning.

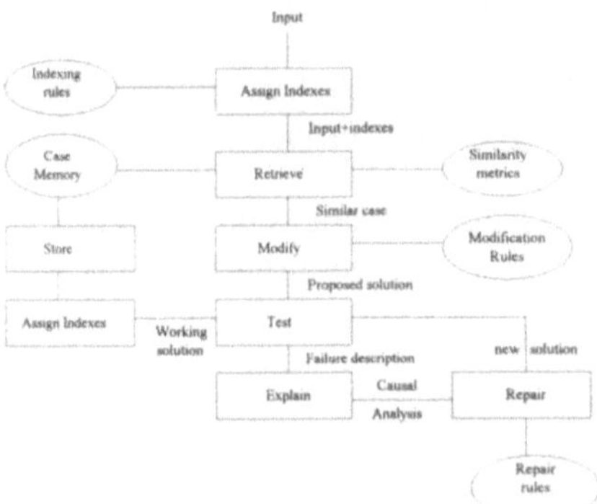

Figure 1.

☐ : processes

❍ : knowledge structures

1. Assign index: the characteristics of a new problem are used as descriptive indices of this problem.
2. Retrieve: the indices and similarity metrics are used to retrieve in the case memory for one similar to the present case.
In chemistry, indices can belong to the "structural display" that symbolizes and expresses numerous analogies and associated metric data.
3. Modification: The past case is modified to include conditions belonging to the current situation, thereby leading to a proposed solution.
4. Test, Explain, Repair: The proposed solution is tested to check its adequacy to the problem. If the test is satisfying, the solution is given to the user, then filed in the case memory. If the test is negative, the cause or causes of the failure must be found and eliminated before retesting.

Studying this process highlights the position of analogy on the problem level between case-based reasoning and structural database management systems. These problems involve indexing information, formulating queries so as to obtain relevant information, similarity metrics, etc. In addition to this beneficial analogy, case-based reasoning enables a flexible and evolving structural DBMS to gradually acquire experience and to become adaptable to the needs of a specific user.

3 SYSTEM ARCHITECTURE AND FUNCTIONING

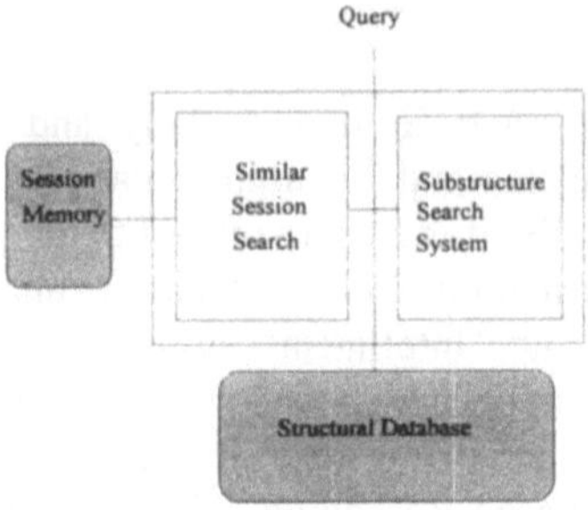

Figure 2. According to this architecture, the substructural search system uses the case-based reasoning much as a chemist would use his own experience to answer a query more satisfactorily. By adding a component part of case-based reasoning (capable of learning) to a substructure search system, we want to achieve a double objective:

i) To adapt a search strategy to a use context by exploiting both the successes and the failures of previous sessions stored in the case memory.
ii) To achieve a long term enrichment of the system know-how by using new knowledge that could then be extracted from the session memory.

The solution we have in mind consists in first elucidating the various possible choices during each phase of the search process (Fig. 3), then adjusting them gradually and progressively all during the use of the system.

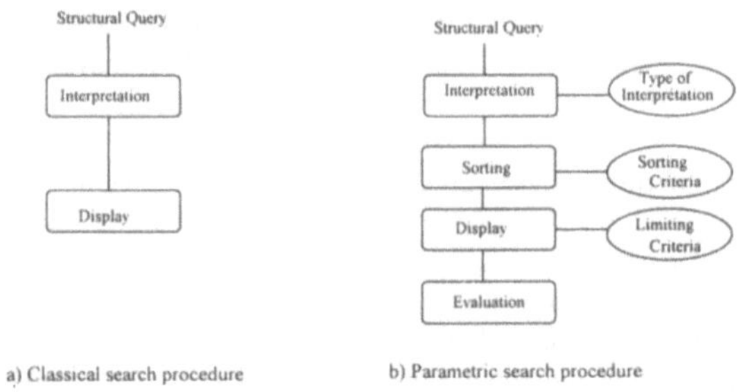

Figure 3. We note that a chemical substructure search system uses conceptual and symbolic expressions of structural knowledge. It is, therefore, a very favorable area of application for testing the ideas evoked below.

3.1 Interpretation

For a substructure search system, to interpret means to select from the database the set of structures with a given structural constraint. This binary restriction: <u>a structural</u>

constraint, verified or not, often obliges the chemist to adapt his queries, thereby rendering the substructure search system rigidly "unintelligent". Perhaps in a search implying a given constraint, the user would be ready to accept some slight differences or even a slightly greater difference without being able to specify the limits of this "slight difference" or "slightly greater difference". In fact, the user does not know in advance what specific criteria of acceptance he can use. He would like the system to suggest those candidates likely to be accepted and then to let him choose among them. The choice of the "best" interpretation in each context is one of the benefits we expect from such experienced managing of a substructure search system.

3.2 Sorting

The selected structures are sorted according to their similarity with the query. The literature proposes many methods for calculating similarity coefficients between a query Q and a structure S. The results of these calculations can be incorporated into the database or in annex as learning elements. According to the case, the similarities are based either on the degree of inclusion of a query Q in the structure S (by identification of the maximal subgraph) or else on the number of structural fragments that the query and the structure have in common. These measures can be combined to allow for several classifications.

3.3 Display

It is during this phase that the choice is made of the number of structures to propose for display and to the user's judgment. There are two possible approaches: either there is a threshold and only those structures whose similarity with the query is greater than this threshold are displayed, or the user is asked to set a maximum of structures, that he accepts, to be visualized.

3.4 Session memory and adapter context

This is the evaluation stage, with the user's help, of the relevance of a search session. Such relevance can be tested by the precision coefficient defined as the number of relevant structures proposed as compared with the total number of structures proposed, or by some heuristic methods, i.e. the user's level of satisfaction with regard to the proposed structures. This stage is very important, for the system learns to improve its performance from one session to another. The great similarity between the characteristics of a search session and a structure led us to propose a session memory organization (case based) identical with the structure base structuration. Just like a structure, a session is characterized by an index and a content. The indices include, in addition to the search primitives, a session relevance index evaluated at the end of each session. The content comprises elements relative to the different search stages, the choices made, the strategy adopted and the list of the chosen compounds.

The search for a session similar to the present is first carried out in the subset of non relevant sessions (those having led to failure), in order to anticipate and prevent this failure. This order of priority, that we can call "learning from failures" corresponds to the first objective we want for our system: to improve search strategy and adapt it to a use context.

CONCLUSION

In this article we describe the main outlines of an approach combining case based reasoning and fuzzy substructure search, the aim of which is to improve the search for information thanks to the use of a session memory. When this memory will be judged sufficiently abundant, we shall have to define a judicious utilization of this experience base so as to enrich the system both in knowledge and in performance. The proximity relations between cases and the metaknowledge that we can excerpt from this memory will enable us to enrich the system's potentialities by browsing tools furnishing reasoned help to the user. With these tools, our system is characterized by openness and flexibility and can generate profile files adapted to specific user needs and conduct chemical diversity searches useful for data mining actions and chemical knowledge creation. These form part of strategies for holding library information. The prototype currently being developed[7], is built around a fuzzy substructure search system based on the latest DARC topological approach which is currently operational in a UNIX environment on a IBM RISC 6000 graphical station.

KEY WORDS

Artificial intelligence, machine learning, analogy reasoning, case based reasoning, substructural search system, fuzzy search, similarity metrics, DARC system.

REFERENCES

[1] Forsyth R., Rada R., Machine Learning: Applications in Expert Systems and Information Retrieval, Ellis Horwood Series in AI (1986)

[2] Chiamella Y., Defude B., A Prototype of an Intelligent System for Information Retrieval: IOTA, Information Processing & Managment System, vol 23 n°4, pp, 285-303 (1987)

[3] Slade S., Case-based Reasoning: a Research Paradigm, AI magazine vol 12, n°1, pp 42-53 (1991)

[4] Chen Z., Enhancing Database Management to Knowledge Base Management: the Role of Information Retrieval Technology, Information Processing & Management system, vol 30, n° 3, pp 419-435 (1994)

[5] Rouvray D.H., The Necessity of Analogies in the Development of Science, J. Chem. Inf. Sci., vol 34, pp 446-452 (1994)

[6] Golging A.R., Rosenbloom P.S., Improving Rule-Based Systems through Case-Based Reasoning. Knowledge Acquisition and Learning (1993)

[7] Work done by SYCOGMA, SA Paris

USING FORMAL SPECIFICATION LANGUAGES TO PROVE DATA MODEL CONSISTENCY

Naïma NAGUI-RAISS

PRISM Laboratory, Versailles University, 45, av. des Etats-Unis, 78000 Versailles, France
E-mail: naima@prism.uvsq.fr

ABSTRACT

To be well defined, data need to be completed by information describing their characteristics. These characteristics represent the data model and are composed of a static part and a dynamic part. Static characteristics describe semantic information such as how data is organized and how some of its elements are related to others (data structures) or under what conditions data are considered sound (integrity constraints). Dynamic characteristics describe the behavior of data with respect to time or to effective surrounding factors such as events. Because of its importance in the data manipulation process, this knowledge must be correctly specified. In order to prove the correction of this specification, formal methods must be used.

This work proposes the use of a formal specification language to specify and prove the consistency of data models.

RESUME

Pour être bien définies les données doivent être complétées par certaines informations qui précisent leurs caractéristiques. Ces caractéristiques représentent le modèle des données et sont composées d'une partie statique et d'une partie dynamique. Les caractéristiques statiques décrivent l'information sémantique telle que la façon dont les données sont organisées et les relations de leurs éléments entre eux (structure des données) ou sous quelles conditions les données doivent être considérées comme saines (contraintes d'intégrité). Les caractéristiques dynamiques décrivent le comportement des données en fonction du temps et des facteurs effectifs de l'environnement tels que les événements. En raison de son importance dans le processus de manipulation des données, cette connaissance doit être correctement spécifiée. Pour prouver l'exactitude de cette spécification des méthodes formelles doivent être utilisées.

Ce travail propose l'utilisation d'un langage formel de spécification pour détailler et démontrer la cohérence du modèle de données.

1 INTRODUCTION

When processing data, it is important to rely on some model. This model is constructed following many observations. These latter can be made by different persons or at different times. For these reasons the specification of the model constructed can be inconsistent. It is then important to prove the consistency of such a model before using it in applications of data processing. In order to prove model consistency, it is necessary to use formal languages to specify this model and formal methods to process this specification. Nevertheless, most engineers are not yet ready to specify their models in languages such as VDM [Jone 80] or B [Abri 93] or make a hand proof of the consistency of their specifications. Support tools are then necessary, not only to help in the specification construction process and the proof process but also in the learning process. These tools must use languages and concepts with which engineers are familiar to introduce the formal notations and methods. Since the Entity-Relationship model [Chen 76] is widely used in the software industry, we have chosen it to be the interface language for the tool presented hereafter.

In this work, we present a support tool for developing data model specifications by using the E/R model and a rule language. This tool helps construct formal specifications containing not only data structures but also knowledge about the data integrity and behavior. The data model is then proved to be consistent. In this paper, we will present the approach adopted for the proposed tool, but we will place more emphasis on the proof process. The formal language used (B), the transformation mechanism from an E/R schema to a B specification and the rule language used to describe data behavior and its mapping onto B notations are presented in more detail in [Nagu 94].

2 THE SPECIFICATION APPROACH

Formal specifications (hereafter FS) have two parts [Abri 94] : a global framework whose construction is time consuming, and a detailed and precise specification of the dynamics. The global framework represents the whole system and the relationships between components of the system. We think that this part is better stated using an E/R model, while the detailed and precise specification must be added by the designer to the initial framework.

Our approach (Fig. 1) consists of generating a B specification from a declarative description of the system and proves the consistency of the specification. This FS tool is integrated in a more complete CASE tool for database application design [Bouz 90]. A graphic and a declarative language interface permit one to define the description of the system in an E/R model. A classical E/R model [Chen 76], extended through the generalization and the aggregation concepts [Smit 77] is used. The E/R schema is transformed into a formal specification in B notation [Abri 94]. The notation and the mapping rules are presented in [Nagu 94]. The dynamics of the system are added through a rule language. They represent the behavior of data, therefore they are described in terms of integrity constraints and management rules. The management rules are described in an ECA (Event Condition Action) formalism [Daya 88].

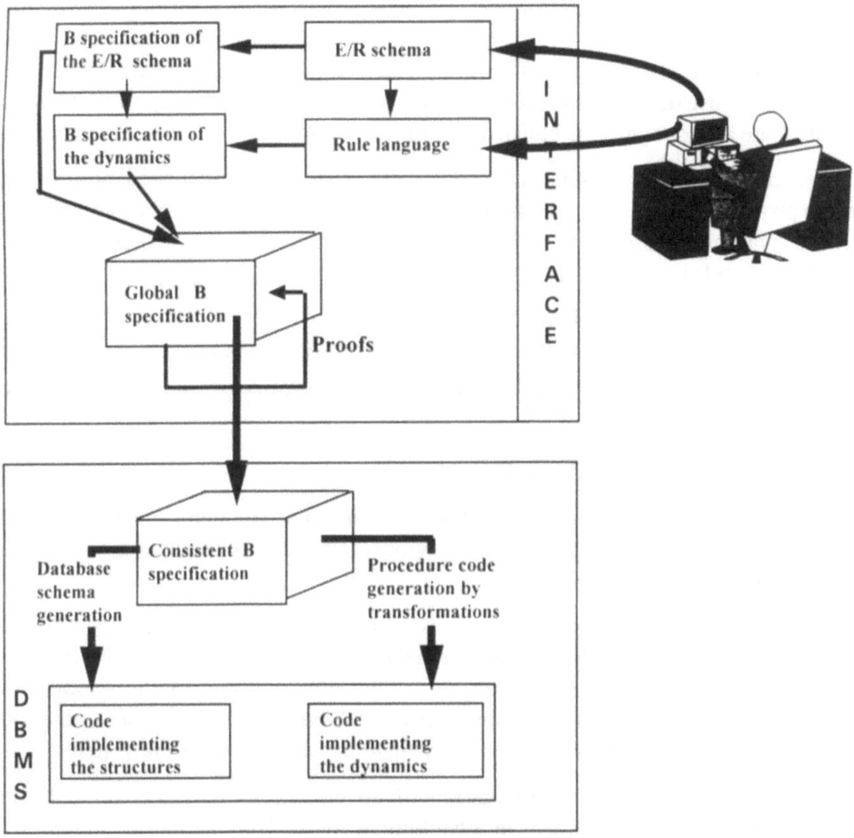

Fig 1 : A formal specification based CASE tool

When the B specification of the data model is generated, the proof process eventually detects data model inconsistency. A consistent specification of the system is transformed into code implementing the database and the operations that allow one to manipulate the data according to the knowledge concerning their behavior. The DBMS target is either a relational or an object oriented one. A transformational approach [Baue 89] is used to generate this code. The initial specification is successively refined using proven transformations, until the last step which produces the executable code. The next section presents the different proofs that can prove the consistency of the model.

3 THE PROOF SYSTEM

A B specification is a set of abstract machines. Each machine has a state, defined by variables, an invariant which describes the static law that the state must verify, and a set of operations that can be performed by this machine. Once a formal specification of the model is constructed, it will be subject to two verification processes: type-checking and proof-obligation [Abri 93]. Type checking verifies that each mathematical construction in the model is well typed. The second process verifies that each data state and operation is consistent with the invariants.

Any predicate involving set-theoretic constructs must be type-checked before even trying to prove it. In order to give an intuitive idea of this checking, let us consider the following example of predicate : this predicate is not well typed, because type(x) _ P(type(x)). Therefore, we will not even try to prove it.

Each machine (unit in the B language) is type-checked. We can not give within this paper all the typing rules, but will provide some representative ones:

- Distinctness conditions: variables and operations must have distinct names.
- The invariant must be well typed.
- Each operation must be well typed: each input parameter must be typed in the pre-condition. Under the previous condition, the substitution must be well typed.

Once an operation has been specified, we have to prove that its specification preserves the invariant. In other words, we have to prove that provided the invariant holds before the operation is executed and if the pre-condition of the operation is satisfied, then the invariant is still true after the substitution defining the operation's effects on the state. The operation does not violate the static laws of the system.

Let us try to explain the proof obligation of an operation "make movement" that allows one to perform a movement in a banking account. This operation expressed in the rule language is as follows:

Make Movement:	**ON**	add movement (acc: Account, am: Real)
	THEN	INSERT acc->Movement (date = date(), amount := am, status := "new")

This is an operation of the machine Banking Account System. The invariant of this machine expresses the integrity constraint that a client's global balance of all his accounts must be positive. The pre-condition of the operation is that the event Add_Movement must be mentioned and that the parameters acc and am must respectively be an "account" and a "real". The substitution is then the insertion of the new movement.

This rule becomes the following operation in the B notation:

$$
\begin{array}{l}
\textbf{\textit{MACHINE}} \\
\textit{BankingAccountSystem} \\
\ldots \\
\textbf{\textit{OPERATIONS}} \\
\textbf{\textit{make\_movement}}\ (acc, am) \\
\quad PRE\ Add\_Movement\ (acc, am) \wedge acc \in account\ \wedge am \in REAL \\
\quad THEN\ mo\ = insert\_Movement\ (date\,())\ || \\
\qquad modify\_amount\_movement\ (mo, am)\ || \\
\qquad modify\_status\_movement\ (mo, "new")\ || \\
\qquad modify\_movement\_account\ (acc, account\_movement\ \leftarrow (acc \mapsto mo)) \\
\quad END
\end{array}
$$

In fact, the rule Make Movement is not consistent with the integrity constraint stated above. The proof system will detect such inconsistency. The user should modify the rule in order to maintain the constraints. The correct rule is:

Make Movement:	**ON**	add movement (acc, am)
	IF	(cli = acc->Has->Client) **AND**
		SUM(cli ->Has ->Account.balance + am) > 0
	THEN	INSERT acc->Movement (date = date(), amount := am, status := "new")

Each operation is proved to maintain the invariant. The invariant contains the typing attributes and the integrity constraints. The proof-obligation verifies if an operation maintains the typing attribute and the integrity constraints.

CONCLUSION

We have presented a support tool to construct formal data model specifications from an E/R model and a rule language. The transformation rules from the declarative language (E/R schema + a rule language) to a B specification have been defined. A B skeleton describing basic operations that maintain local integrity constraints is generated for each entity or relationship.

The objective of this tool is to introduce formal specification languages in the industry and familiarize engineers with formal methods. Complete consistency proofs can be executed over the global specification. These proofs can detect typing errors or contradictions in the description of data behavior. These contradictions are represented either by inconsistency in some machine's invariant or the fact that operations do not maintain the system invariant (typing attribute and integrity constraints). Type checking can detect errors in the static E/R model. For instance, cycles in the generalization hierarchy can be detected.

KEYWORDS

Data, data structure, data behavior, formal software specification, formal specification language, CASE tools, consistency proof, integrity constraints, ECA rules.

REFERENCES

[Abri 93] Abrial J.R., The B method, to be edited, 26, rue des Plantes, Paris

[Abri 94] Abrial J.R., The B notation, A CNAM course Notes, Paris, January-June (1994)

[Baue 89] Bauer F.L., Möller B., Partsch H., Pepper P., Formal program construction by transformations - Computer Aided, Intuition Guided programming, IEEE Trans. Soft. Eng., Vol 15, N°2, Feb (1989)

[Bouz 90] Bouzeghoub M., Metais E., Leborgne L., Hazi F., A Design Tool for Object Oriented Databases, Proceed. of the 2nd Conference on Advanced Systems Engineering (CAISE90), Stockholm, Sweden, May (1990)

[Chen 76] Chen P.P., The Entity-Relationship Model, Toward a Unified View of Data, ACM Transactions on Database Systems, Vol.1, N°1, pp. 9-36, March (1976)

[Daya 88] Dayal U. et al, The HIPAC Project: Combining Active Databases and Timing Constraints, SIGMOD Record, Vol.17, N°.1, March (1988)

[Jone 80] Jones C., Software development: a rigorous approach, London, Prentice-Hall (1980)

[Nagu 94] Nagui-Raïss N., A formal Software Specification Tool Using The Entity-Relatioship Model, to be published in the Proceedings of the Thirteen Inter. Conf. on the E-R approach, Manchester, Dec.(1994)

[Smit 77] Smith J.M., Smith D.C.P., Database Abstractions: Aggregation and Generalization, ACM Transactions on Database Systems, Vol.2, N°2, pp.105-133, June (1977)

CHEMICAL ABSTRACTS INDEX NAMES: AUTOMATIC GENERATION

Shengang YUAN, Chongzhi ZHENG and Jianhua YAO

Laboratory of Computer Chemistry, Shanghai Institute of Organic Chemistry, Chinese Academy of Sciences, Shanghai 200032, China

ABSTRACT

The complexity for assigning the systematic name (IUPAC system or CA index name) to a structure led to the development of computer software to support the automatic generation of nomenclature. The Chemical Abstracts (CA) index names are very useful for finding various information through CA. We designed a structure naming algorithm whose core is the unic Bridge-E_Block approach. This approach dissects the structure into E_Blocks and converts it into E_Block Bridge Tree (EBBT) in which each node corresponds to a namable structural fragment. This approach makes it easier to identify namable structural fragments, parent selection and substituent determination. Based on the approaches proposed in this paper a naming system with high performance could be developed.

RESUME

La complexité de l'assignation d'un nom systématique (système IUPAC ou nom indexé CAS) à une structure a conduit au développement de logiciels pour aborder la génération automatique de la nomenclature. Les noms indexés du Chemical Abstracts (CA) sont très utiles pour chercher des informations variées dans CA. Nous avons conçu un algorithme de génération d'un nom structural dont le centre est du type unic Bridge-E_Block. Cette approche dissèque la structure en Bloc E et la convertit en un arbre de Bridge-E_Block (EBBT) dans lequel chaque noeud correspond à un fragment structural qui peut recevoir un nom. Cette approche facilite l'identification des fragments structuraux à "nom", la sélection de parents et de substituants. Basé sur les approches proposées dans cet article un système de nomenclature de haute performance pourrait être développé.

1 INTRODUCTION

The Chemical Abstracts (CA) index names are very useful in chemical communication, particularly as keys for finding information through CA. Their greatest inconvenience is the amount of time-consuming effort involved in generating the correct name. It is recognized that a computer software system to support the automatic transformation of connection table into CA index name is necessary[1], and similar systems aimed at generating IUPAC nomenclature have appeared[2,3]. For broad usefulness such a system should not be limited to specific classes of compounds, but be a general purpose system for the unambiguous translation of structure diagrams of organic compounds into CA index names. A complete naming system which we developed recently consists of six major phases.

1 Structure Input
2 Structure Preprocessing
3 Identification of Namable Structural Fragments
4 Ring Analysis
5 Parent Selection and Substituent Determination
6 Name Assembly

2 Description of the Major Phases

2.1 Structure input

The first step in naming a structure is to accept the structure of the compound being named. This structure is prepared and accepted by a graphic interface and stored in the form of a connection table, processed by successive phases to generate the name. Commercial structure input software, e.g. ISIS is now available. The system was designed to accept the structure by its own interface and also by reading from commercial software files

2.2 Structure Preprocessing

Since structures for some kinds of compounds are not unambiguous, different structures can represent an identical compound, and different input systems store structures in different kinds of structure representation. The input structures should first be converted into a form to which nomenclature rules can be applied.

2.3 Identification of namable structural fragments (ring systems, functional groups and chains)

The fundamental principle of systematic nomenclatures (IUPAC system, CA, DARC/QUESTEL or Beilstein systems) is (1) to produce a number of namable structural fragments (ring systems, functional groups and substituents) in the structure of the compound being named according to the nomenclature rules; and (2) to select the name of the fragment with highest seniority (parent ring system or the longest straight chain) as parent name, others are named as substituents on the parent. The parent name is combined with substituent names, locants, and letters designating stereochemistry. Thus, it is logical that the first step for naming a compound by a computer system is to dissect the structure into namable structural fragments, namely ring systems, functional groups (substituents) and chains.

Bridge-E_Block Approach.

Because the identification of a substructure in structure involves many time-consuming combinatorial problems, the computation time or space needed is exponentially rather than linearly dependent on the number of atoms in the structure. To enhance identification efficiency, one must try to diminish the number of atoms in the structure during identification and to use efficient combinatorial algorithms.

Contrary to the usual approach, i.e. to identify functional groups and ring systems by means of an atom-by-atom matching and algorithm based on the Smallest Set of the Smallest Rings (SSSR) against the whole structure respectively, in our work, we used a newly developed algorithm for identifying functional groups and ring systems based on a bridge-E_block approach, i.e. to find the bridges in the structure first and dissect it into E_blocks by removing the bridges, to identify the functional groups (substituents) and ring systems in acyclic and cyclic E_blocks respectively. This new approach distinguishes cyclic from acyclic E_block by their characteristics. If the number of atoms is less than or equal to that of bonds in an E_block, it is a cyclic E_block; otherwise it is called an acyclic E_block. The identification of functional groups requires a significant amount of subgraph isomorphism determination. In this approach, the structure is dissected into E_blocks which are distinguished into ring system, chain and link before the recognition of functional group, ring system and chain. Thus, functional group identification is only needed among acyclic E_blocks and not among cyclic E_blocks. Similarly, the identification of most parents can be carried out only in cyclic E_blocks. With such an approach identification is carried out in the part of the structure which contains less atoms than the whole, and a high efficiency can be obtained.

Efficient Subgraph Isomorphism Algorithm

Among several algorithms proposed for the subgraph isomorphism problem, Ullman's [4] seems suited to identifying functional groups, as some researchers show it to be faster than the others [5]. We apply Ullman's subgraph isomorphism algorithm to finding whether a given functional group is a substructure of an appropriate E_block. By modifying Ullman's algorithm an improved subgraph isomorphism algorithm was developed to identify the functional group. The main improvement minimizes the number of '1' in initial matrix M0 by maximal use of the characteristics of the first layer of each atom (connectivities, bond types, atom nature, etc.).

2.4 Ring analysis

Ring system naming and ring atom numbering (assigning ring locants) are the most complex and difficult problems in structure nomenclature. Some programs have been reported [6,7]; the ability to name arbitrary ring systems has not yet been envisaged. To obtain the trivial name of a ring system, a dictionary with most common ring systems is searched. For a ring system not in the dictionary, the system asks the user to input its trivial name. To improve atom-by-atom matching with each entry of a predefined dictionary, to access the entry, its trivial name and locants quickly, a sophisticated indexing system for matching between ring system and entries must be used. It can be built by a thorough analysis of the ring system.

Our first step was to generate all the possible rings forming the "all-ring set" of the system. The information in this set is exhaustive and useful for ring analysis. The algorithm for generating the all-ring set followed that proposed by Balaban[8].

An efficient hashing system can be developed with information contained in the all-ring set. To generate the hashing code one needs ring sizes, the number of all-rings, their elemental component and multiple bonds. Thus, only a few entries need be compared with the ring system by an atom-by-atom match. Once the chains, functional groups and ring systems are recognized, a parent fragment can be selected according to nomenclature rules.

2.5 Parent selection and substituent determination

The parent structure selection is to find the structural fragment to be used as parent in name generation among the recognized namable fragments by applying the nomenclature seniorities, and the remaining structural fragments become substituents on the selected parent structure. Once the parent structure and substituent fragments have been determined, the parent name, substituent names and locants can be generated. We do this at the EBBT level where each node corresponds to a namable structural fragment. At this EBBT level the parent node is selected by applying 13 general rules used by nomenclature specialists at CAS[1]. In some cases, a new node must be created to add into the EBBT, e.g. a combined ring system is encountered; in other cases some nodes in the EBBT should be merged, e.g. a ring assembly is detected. After we traverse the EBBT starting from the parent node (root) and reorganize (create and/or eliminate the nodes), a new tree called nomenclature tree is generated.

Generation of EBBT

After bridges and E_blocks have been identified, EBBT is generated simply by applying a connected component finding algorithm. The EBBT can be processed as an ordinary tree (graph), the only differences between an EBBT and a graph are that in EBBT each node represents an E_block and the edge a bridge.

Ring Assembly Identification

A ring assembly in CAS index nomenclature is a linear "chain" of two or more identical rings joined by acyclic single bonds, not necessarily at equivalent positions. In EBBT every node represents an acyclic or cyclic E_block which is easily identified by its order and size. Any identical cyclic E_block joined by acyclic single bonds is a ring assembly. Such joined E_blocks are found easily at the EBBT level.

2.6 Name assembly

To generate the compound name, we traverse the nomenclature tree starting from the parent node. As each node is visited a corresponding name will be found and these names, or character strings, are organized to form the CA index name according to the assembly rules. The CA index name is the concatenation of these strings, inserting punctuation elements, locants, multiplier, prefixes in an appropriate order. Before the resulting name is output, it must be checked by several operations (e.g. substituent multiplication, punctuation, vowel deletion, suppression of unnecessary locants, superscript and italic string placement, number of square brackets, etc.).

CONCLUSION

It has been recognized that naming a compound requires much time-consuming effort. In the past years several automatic generation systems for IUPAC systematic nomenclature appeared. Our reasons for developing an automatic system of generating CA index names are: (1) a unique CA index name can be obtained by applying CA nomenclature rules and (2) a CA index name is a key for finding information through CA, particularly when on-line structure/substructure searching is not available, as in China at present. Based on the approaches presented in this paper a system capable of naming usual compounds was developed on a MICRO VAX 3300 in C language. We are attempting to complete all naming capabilities and to transfer the system onto a PC platform.

ACKNOWLEDGMENT

The development of the Automatic Generation of Chemical Abstracts Index Names is supported by the National Natural Science Foundation of China. The authors and the Laboratory of Computer Chemistry (Shanghai) gratefully acknowledge this support. The discussion with CAS specialists, especially B.J. Bossenbroek, J. Mockus and P.M. Giles, Jr., in 1991 was helpful for this work. They and others whose names could not be noted here are also gratefully acknowledged.

APPENDIX - DEFINITIONS

A component of a graph G is a connected subgraph of G not properly contained in any other connected subgraph of G. The number of connected components of a graph is denoted by k(G).

An edge e of a graph G is called **bridge** if the removal of e increases the number of connected components of the graph, k(G-e)>k(G). If e is a bridge of G, then it is immediately evident that k(G-e)=k(G)+1. A **cycle edge** is an edge that lies on a cycle.

A cycle edge of a graph G is an edge that is not a bridge of G. A bridge incident with an end-vertex is called a **terminal edge**. A **cycle-vertex** is a vertex that lies on a cycle. A bridge incident with at least one cycle-vertex is called a **link**.

A connected graph G is nonseparable if G has no cut vertex. A nontrivial connected graph with no cut-vertices is called a **block**.

In this work, we are concerned with the blocks induced by removing all bridges. These are called **E_block** and denoted by B'. There are only two categories of these: (1) order p = 1; and (2) order p ≥ 3. Moreover, the latter, E_block, is called **cyclic E_block** while the former is called **one-vertex E_block**. We join one-vertex E_blocks by restoring the removed bridges between them and call the joined E_blocks **acyclic E_blocks**.

An acyclic E_block is called **terminal(chain) E_Block** if there is at least one end-vertex; otherwise it is called **link E_block**. An **E_Block-Bridge Tree (EBBT)** is defined for a given graph G as follows: EBBT=(B', BR) is an E_block-Bridge graph of G if: (1)B' is a set which is the union of the acyclic and cyclic E_block sets of the graph G and (2)BR is a set of bridges whose two vertices belong to different E_blocks of B'.

To meet chemists' conventions, we often use the terms with more chemical significance: chain, ring system, terminal chain and bond rather than acyclic E_block, cyclic E_block, terminal E_block and link E_block.

KEYWORDS

Systematic name, CA index name, structure nomenclature, namable structural fragment, parent name selection, substituent determination, ring analysis.

REFERENCES

[1] Mockus J., Isenberg A.C., Vander Stouw G.G., J. Chem. Inf. Comput. Sci., 21, 183-195 (1981)
[2] Meyer D.E., Gould, S.R., Am. Lab., 20(11), 92 (1988)
[3] Wisniewski J.L., J. Chem. Inf. Comput. Sci, 30, 324 (1990)
[4] Ullman J.R., J. Assoc. Comput. Mach., 23, 31-42 (1976)
[5] Brint A.T., Mitchell E., Willett, P., Substructure searching in files of three-dimensional chemical structures, in W.A. Warr (ed.) Chemical structure, The International Language of Chemistry.
[6] Rucker G., Rucker C., Chimia, 44, 116-120 (1990)
[7] Van Binnendyk D., Mackay A.C., Can. J. Chem, 51, 718-723 (1973)
[8] Balaban A.T., J. Comput. Chem., 6, 316-329 (1985)

An International Database of Nuclear Concrete Structures Ageing

***Claude SENI*[1] *and Leonid IANKO*[2]**

1 Atomic Energy of Canada, Ltd.,AECL,Mississauga,Ontario, Canada
2 International Atomic Energy Agency, Division of Nuclear Power, Vienna, Austria

Abstract

Under the initiative and auspices of the International Atomic Energy Agency (IAEA) in Vienna and with the technical support of Atomic Energy of Canada-Ltd. (AECL), an International Database of Concrete Structures Ageing is being created. This paper discusses the background, mechanisms and problems associated with its creation, its use and benefits, i.e. the activities involved with the collection of the data, validation and processing into a standardized reporting system in order to make the information reliable and interchangeable among nuclear utilities, the preservation of the confidentiality of the data received from the utilities, the access to the data by the utilities, etc. The Database structure and imports from other existing Databases are also discussed.

Resume

A l'initiative et sous les auspices de l'Agence Internationale de l'Energie Atomique (AIEA) de Vienne et avec le soutien technique de l'Energie Atomique du Canada-S.A. (EACL), on crée une base de données (BD) sur le vieillissement des structures de béton. Dans cet article, on discute de l'historique, des mécanismes et des problèmes associés à sa création, à son utilisation et à son utilité, voire les activités associées à la collecte, à la validation et au traitement des données pour aboutir à un système standardisé de consultation. L'objectif central est d'assurer une information sûre et échangeable entre les centrales nucléaires tout en préservant la confidentialité des données des fournisseurs et leur accès aux données, etc. La structure de la BD et les contributions d'autres BD existantes sont discutées.

1 Introduction: The Need and Novelty of the Database

The ageing of nuclear structures is a process which is of special interest because of the extended service life of these structures and the potential impact their deterioration can have upon the safety and reliability of the operation of nuclear facilities. Such structures include nuclear power plant concrete containment designed to separate the reactor and other systems from the outside environment, internal support structures, spent fuel bays, etc.

Although a large variety of databases exist or are being created in the scientific world, in the field of concrete and concrete structures, they address material properties in general and do not focus on material performance.

In 1992 an initiative was taken by the International Atomic Energy Agency (IAEA) in Vienna with the technical collaboration of the Atomic Energy of Canada Ltd. (AECL) to create a database dedicated to gathering information on how nuclear concrete structures perform with time. This time the focus is not on properties but on performance, i.e. the effect of AGEING and its impact on the way the structures perform with time under service conditions. Such a database can be a valuable tool for the design engineer to select the most appropriate materials and for the operating engineer to monitor the performance of a structure and plan maintenance programs or repairs. Considering the service life of concrete nuclear structures which may have to exceed 100 years for containment or 500 years for underground repositories, the importance of this database becomes even more obvious. Moreover, while the interest in the ageing of ordinary concrete structures is mostly STRUCTURAL, i.e. deterioration of the structural capacity to withstand loads, for nuclear structures a concern also exists for FUNCTIONAL ageing, i.e. deterioration of their leak tightness. The latter, negligible in the case of buildings, bridges and even concrete tanks and containers, is of utmost importance for the concrete containment of a nuclear power plant, for reasons explained earlier, and thus is addressed in this database.

To achieve the above objective, the IAEA have launched a survey of all nuclear power plants around the world which will ensure that the database will incorporate the latest, world wide experience in the field.

2 OBJECTIVES

The objectives of the database will be to:

2.1 Obtain an understanding of the ageing mechanism and impact

The structural ageing mechanism is sufficiently well known, but the impact upon various structures will depend on the way they were designed, i.e. the governing design criteria. For ordinary structures, like buildings or bridges, the concrete members are designed allowing the concrete in the tension zone to crack in order for the reinforcing to become fully effective, while cracks are not acceptable on some of the concrete nuclear structures even if the structural integrity is ensured. This will create different conditions for the ageing agents to act and to affect the component performance.

In addition, the functional ageing mechanism, specific to nuclear components like concrete containment or spent fuel bays, resulting in the gradual loss of leak tightness, needs to be better understood.

2.2 Quantify the ageing effect vs. time

The database will permit the establishment of ageing curves for each component, which could then be used to assess their state of health and predict their residual service life.

2.3 Establish average values and trends

Disposing of a larger number of data for the same type of ageing factor and component, it will be possible to define ageing curves and trend values. These could then be compared with those of a specific component and could determine ahead of time if the component is ageing more rapidly, if there is a problem, its nature and how to address it.

3 METHODOLOGY

3.1 Technical specification

A technical specification was developed by IAEA with a panel of experts, setting the ground rules for data collection, selection, processing, use, quality assurance, etc.

3.2 Statistics

The database will be based upon the collection of a sufficiently large number of data to make possible the establishment of average figures according to statistical practices.

3.3 Three party collaboration

The undertaking of this task involves three parties:

- The **Contributors,** which are the nuclear utilities that supply the data,
- The **Custodian**, the organization which keeps and manages the database,
- The **Participants**, the organizations approved to use the database.

3.4 Agreement

Specific plant data may be regarded as sensitive by the Contributors. Thus the confidentiality of the source has to be ensured.

Rules to this effect will be documented and form part of an Agreement, to be signed and observed by all parties.

4 Source of Data

Most of the data will be provided by the nuclear utilities based upon the world wide survey initiated by IAEA. This information about structures operating under actual in-service conditions will be more realistic than data generated under laboratory conditions since the laboratory data is generally based upon accelerated ageing tests which could distort the results.

Where needed, these data will be supplemented with R&D data. Additional data will also be sought, if applicable, from other existing databases of nuclear or non-nuclear facilities.

5 Selection Criteria

IAEA's methodology of ageing management which will be followed consists of selecting data relating to the maintenance of physical barriers important to the safety of operation.

In the case of concrete structures this represents structures with a containment/confinement role.

The above safety criterion and economic considerations were the basis for component classification into four categories:

(i) those important to safety/reliability and which are irreplaceable

(ii) as above and difficult or expensive to replace

(iii) as above and replaceable

(iv) others.

6 Database Specification

The database specification elaborated for this purpose addresses the following aspects:

6.1 Data collection and qualification

Since data for inclusion in the database will come from various countries, rules are provided for the preparation of the data by the Contributor and the qualification by the Custodian, i.e. format, standards and unit to be used, etc, in order to achieve consistency, thus international acceptance.

A level of confidence will be assigned to each data at the time of qualification by the Custodian.

6.2 The database format

In the process of developing the database, three levels of databases will be produced:

1) raw data, as received from the Contributors,
2) data evaluated and qualified by the Custodian and
3) data ready for use.

6.3 Data verification, processing and validation

Rules are defined for the Custodian who will receive raw data and will have to convert them into their final form for inclusion in the database.

6.4 Database usage

Rules and conditions are provided for the right of accessing and using the database by the Participants, in order to ensure the confidentiality of the source and prevent misuse. The Participants will have to sign an Agreement documenting these aspects.

6.5 Database maintenance

Rules are provided for the Custodian on how to keep, expand and update the database. A form of feed-back will be defined between the Custodian and the Participants.

Other chapters of the specification cover rules for the quality assurance, data distribution and protection, for the software and hardware and the user's manual.

7 AGEING: SPECIFIC DATA REQUIREMENTS

Consistent with the prime objective of the database i.e. to address the process of ageing, a number of parameters have been selected which could best define the process and its impact on the concrete and concrete structures. This selection was based upon an understanding of the degradation agents and mechanism, an extensive literature search and results obtained by researchers.

The following types of degradation were selected for the first reporting phase: weathering, leaching, chemical attack, corrosion, irradiation and high temperature.

As mechanisms of ageing, the following were selected: carbonation, chloride penetration, cracking, permeability, creep, shrinkage, freeze/thaw and the alkali aggregate reaction.

Beside the collection of data describing deterioration, data will also be collected and incorporated in the database regarding methods used to detect deterioration and the rehabilitation procedures, i.e. inspection/monitoring programs, repair methods, protection/mitigation methods applied and results obtained. This will help define the most effective method for each case and also provide a reference procedure.

CONCLUSION

Once the Agreement is signed between the users (Participants) and the Custodian of the Database, each Participant will be able to electronically access the database from a remote location, through an existing network which will connect all nuclear stations, e.g. the existing WANO network.

The data will be used to establish ageing curves and make predictions regarding the expected performance of the concrete components, e.g. the leak tightness of the containment, or the residual service life.

It will also be possible to anticipate deterioration due to ageing, and allow a utility to plan ahead of time the necessary repairs, minimizing cost and interruption of operation.

The database, which collects world wide experience, will provide the most appropriate and proven repair methods for each case.

It will permit a comparison of component ageing with average values and decide on actions to be taken to mitigate the ageing effect.

Finally, the state of health and expected residual life which could be worked out with the help of the database, could be used to support the renewal of a plant operating license.

KEYWORDS

Concrete, nuclear, database, ageing.

INDEX

Volumes Published in the Series
Data and Knowledge in a Changing World

The Information Revolution: Impact on Science and Technology
Edited by J.-E. Dubois and N. Gershon

Modeling Complex Data for Creating Information
Edited by J.-E. Dubois and N. Gershon

MIX
Papier aus verantwortungsvollen Quellen
Paper from responsible sources
FSC® C105338

If you have any concerns about our products,
you can contact us on
ProductSafety@springernature.com

In case Publisher is established outside the EU,
the EU authorized representative is:
Springer Nature Customer Service Center GmbH
Europaplatz 3, 69115 Heidelberg, Germany

Printed by Libri Plureos GmbH
in Hamburg, Germany